Electrical motor products

Related titles:

Materials for energy efficiency and thermal comfort in buildings
(ISBN 978-1-84569-526-2)
Materials for energy efficiency and thermal comfort in buildings critically reviews the advanced building materials applicable for improving the built environment. Almost half of the total energy produced in the developed world is inefficiently used to heat, cool, ventilate and control humidity in buildings, to meet the increasingly high thermal comfort levels demanded by occupants. The utilisation of advanced materials and passive technologies in buildings would substantially reduce the energy demand and improve the environmental impact and carbon footprint of building stock worldwide. Part one reviews both fundamental building physics and occupant comfort in buildings, from heat and mass transport, hygrothermal behaviour, and ventilation, on to thermal comfort and health and safety requirements. Part two details the development of advanced materials and sustainable technologies for application in buildings, beginning with a review of lifecycle assessment and environmental profiling of materials. The section moves on to review thermal insulation materials, materials for heat and moisture control, and heat energy storage and passive cooling technologies. Part two concludes with coverage of modern methods of construction, roofing design and technology, and benchmarking of façades for optimised building thermal performance. Finally, part three reviews the application of advanced materials, design and technologies in a range of existing and new building types, including domestic, commercial and high-performance buildings, and buildings in hot and tropical climates.

Generating power at high efficiency: Combined cycle technology for sustainable energy production
(ISBN 978-1-84569-433-3)
A secure energy supply is essential to our way of life. However, diminishing fossil fuel reserves, the slow development of renewable energy generation and concerns about climate change have all focused attention on maximising the efficiency of conventional power stations. Using examples drawn from plants around the world, this important book reviews key developments in combined cycle technology in helping to ensure sustainable energy production. It presents some of the issues which are shaping the energy system of the future and discusses how the combined cycle can evolve to meet the different needs of a future market driven by a greater contribution of base-loaded nuclear and renewable energy sources, mainly wind and solar power. Based on a wealth of experience, and drawing on a range of practical examples, Generating power at high efficiency will be a valuable reference for all those wishing to improve the efficiency of the power industry.

Electrical motor products

International energy-efficiency standards and testing methods

JIANFENG YU, TING ZHANG,
JIANMING QIAN

WP
WOODHEAD
PUBLISHING

Oxford Cambridge Philadelphia New Delhi

Published by Woodhead Publishing Limited,
80 High Street, Sawston, Cambridge CB22 3HJ, UK
www.woodheadpublishing.com

Woodhead Publishing, 1518 Walnut Street, Suite 1100, Philadelphia,
PA 19102-3406, USA

Woodhead Publishing India Private Limited, G-2, Vardaan House,
7/28 Ansari Road, Daryaganj, New Delhi – 110002, India
www.woodheadpublishingindia.com

First published 2011

British Library Cataloguing in Publication Data
A catalogue record for this book is available from the British Library.

Library of Congress Control Number: 2011919189

Woodhead Publishing ISBN: 978-0-08-101710-4 (print)
ISBN: 978-0-85709-381-3 (online)

Typeset by RefineCatch Limited, Bungay, Suffolk, UK

Contents

List of figures and tables

Figures

Tables

Acknowledgements

The authors gratefully appreciate the following people for their help in preparing the book manuscript. Professor Huixin Yuan (Changzhou University), Professor Yuehong Yin (Shanghai Jiao Tong University), Professor Zhicheng Ji (Jiangnan University), Master of Science Zhen Chang (University of Ulm, Germany), Assistant Professor Ganyun Lv (Zhejiang Normal University), Assistant Professor Yanchu Xia (Shanghai Second Polytechnic University), Ing. Simona Hrinikova (Hewlett-Packard, Sro., Slovakia), Assistant Professor Yanxia Shen (Jiangnan University), Dr Jiansong Chen, Jianfeng Hu (Jiangsu Energy Efficiency Testing Center), Youli Zhao (Science and Technology Bureau of Xishan District, Wuxi), Yuhua Zheng (Jiangsu Inspection and Quarantine Mechanical and Electrical Products Testing Centre), Fenggui Huang (Wuxi Hongyuan Textile Machinery Co. Ltd), Dr Jerry Wang (University of Illinois at Urbana-Champaign), and Master of Science Bill Zhang (University of Oulu, Finland).

The authors would also like to thank Dr Glyn Jones and Jonathan Davis at Chandos Publishing for their great support and encouragement during the publication of this book.

About the authors

Jianfeng Yu is currently director of the Technical Development Department at Jiangsu Inspection and Quarantine Mechanical and Electrical Products Testing Centre. He is responsible for researching new testing technology for mechanical and electrical products.

Following a BSc and MSc in mechanical engineering at Wuxi University of Light Industry (now named Jiangnan University), a PhD in mechatronic engineering at Shanghai Jiao Tong University in 2004 and two years' work as a director technologist of the Research and Development Department at the Wuxi Diesel Engine Works, FAW Jiefang Automotive Company Ltd, Jianfeng started his career as a researcher at Jiangsu Inspection and Quarantine Mechanical and Electrical Products Testing Centre in 2006. In 2007, he was appointed as director of the Technical Development Department at Jiangsu Inspection and Quarantine Mechanical and Electrical Products Testing Centre. From 2008 to 2009, he was involved in setting up the passive safety of automotive parts testing lab and he was responsible for introducing the computer-controlled ServoSled system from the SESA company. Since 2010, Jianfeng has been building the testing lab for the transducers applied in the Internet or wireless sensor networks; the reliability and the electromagnetic capability of the transducers are tested in the lab.

Jianfeng is a member of Science and Technical Committee of Wuxi Entry-Exit Inspection and Quarantine Bureau. He is also the secretary of the Technical Committee of Mechanical and Electrical Engineering, China Inspection and Quarantine Standardization Association. From 2008 to 2010, he was a coordinator of the research project 'Energy efficiency technical measurements for imported and exported energy used products' supported by the General Administration of Quality Supervision, Inspection and Quarantine of the People's Republic of China (Grant No. 2008IK260); 26 energy used products were researched in the project, including electric motors, pumps and fans.

Furthermore, he has presided over and participated in more than 12 national-level, provincial-level and enterprise-funded research projects. Jianfeng owns four patents and has published more than 30 scientific papers in journals and conferences. He is also a reviewer of the *International Journal of Advanced Manufacturing Technology*.

Jianfeng Yu may be contacted at:

Jiangsu Inspection and Quarantine Mechanical and Electrical Products Testing Centre No. 328, Yanxin Road
Huishan District,
Wuxi, 214174, Jiangsu
China
E-mail: *robotmcu@126.com*

Ting Zhang is currently director of the Wuxi Entry-Exit Inspection and Quarantine Bureau, a role he was appointed to in April 2009. Prior to this he was director of Changshu Entry-Exit Inspection and Quarantine Bureau and director of Jiangyin Entry-Exit Inspection and Quarantine Bureau. From 2007 to 2010, he was one of coordinators of the scientific research project 'Management of energy efficiency for home appliances and its technical measures', funded by the Ministry of Science and Technology of the People's Republic of China. From 2008 to 2010, he was responsible for the research project 'Energy efficiency technical measurements for imported and exported energy used products', supported by the General Administration of Quality Supervision, Inspection and Quarantine of the People's Republic of China (Grant No. 2008IK260). He has rich experience and innovation in the field of import and export commodities inspection and quarantine.

Ting Zhang may be contacted at:

Wuxi Entry-Exit Inspection and Quarantine Bureau
No.10, Huaxia Middle Road
Xishan District
Wuxi, 214101, Jiangsu
China
E-mail: *wuxiciq@126.com*

Jianming Qian is currently director of the Jiangsu Inspection and Quarantine Mechanical and Electrical Products Testing Centre. He graduated from the University of International Business and Economics in 1991. He is an excellent technical talent and academic leader. He has devoted himself to inspecting mechanical and electrical products for more than 30 years and is familiar with the international electrical safety standards for mechanical and electrical products. Under his leadership (1995–2011), the testing centre has made rapid developments. He holds four patents, including for a UV sterilization device, multi-function lamps and an electronic newspaper reader. He has designed 50 testing devices, which have been applied in the field testing work. He won second and third prize for 'Rejuvenating the quality inspection through science and technology', issued by the General Administration of Quality Supervision, Inspection and Quarantine of the People's Republic of China.

His life motto is: 'Be good at watching and thinking, broadly establishing the range of knowledge. A combination of theory and practice is the way to win success.'

Jianming Qian may be contacted at:

Jiangsu Inspection and Quarantine Mechanical and Electrical Products Testing Centre
No.328, Yanxin Road
Huishan District
Wuxi, 214174, Jiangsu
China
E-mail: *wuxirobot@126.com*

1

Introduction: the importance of energy-efficiency for electrical motor products

Abstract: This chapter introduces the application of electrical motor products. The chapter provides the history of the electric motor, pump and fan; it describes the current situation for the energy-efficiency of electrical motor products and points out the importance of energy-efficiency for electrical motor products against the low-carbon economy background.

Key words: electrical motor products, energy-efficiency.

1.1 The application of electrical motor products

Global climate change has led to great concerns by global society. The negative impact of greenhouse gases (GHG) on climate change is clear now. The reduction of GHG can be achieved either by reducing total energy consumption or by increasing energy-efficiency. Improving energy-efficiency has received considerable attention as a way of reducing GHG emissions and alleviating energy shortage. In recent decades, much work has been done to improve the energy-efficiency of various equipment and appliances, including:

- boilers and water heaters;
- heating, ventilation and air-conditioning systems for commercial and residential buildings;
- cooking equipment;
- household appliances such as refrigerators, washing machines, dishwashers and other small appliances;
- lighting;
- electronic appliances (televisions, computers, digital video recorders);
- office equipment (copiers, fax machines);
- miscellaneous electronic devices (power supplies, stand-by power).

As for developed countries, the Group of Eight (G8: Canada, France, Germany, Italy, Japan, Russia, the United Kingdom and the United States) has achieved great progress in the energy-efficiency field. The G8 represents a significant share of world primary energy consumption (about 46 per cent) (United Nations Foundation, 2007). By working effectively with the +5 nations (China, India, Brazil, Mexico and South Africa), the G8 countries can spur efficiency improvements in economies that together consume nearly 70 per cent of global primary energy. As for developing countries, for example, China, Brazil, India, Mexico and South Africa have made effective energy-efficient policies and measures to accomplish the low-carbon emission target.

With regard to electrical motor products, the International Energy Agency (IEA) regards electric motor, pump, fan and compressor as industrial motors. Electrical motor products have been widely applied in industry, consuming roughly 60 per cent of electricity in industry, or 40 per cent of global electricity, as IEA surveyed. Electric motors supply mechanical energy through the electromagnetic transformation principle,

Figure 1.1 Disaggregation of motor electricity consumption by end-use in the industrial sector (Used with permission of Prof. Anibal T. de Almeida)

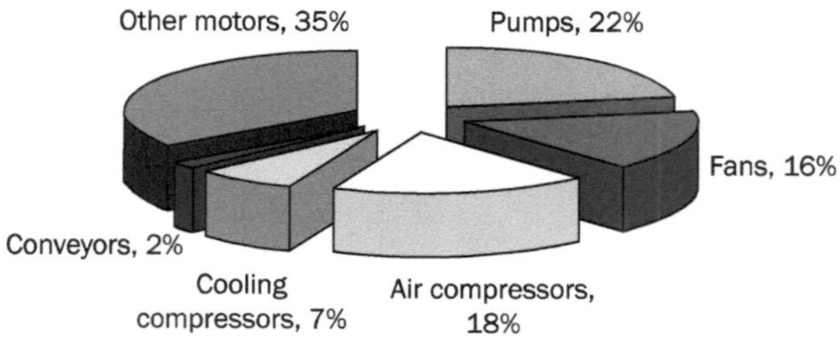

while pumps and fans supply fluid energy. Electrical motor products are widely applied in industry, agriculture, and so on. Electric motors, pumps and fans are by far the most important loads in industry, accounting for 73 per cent (Figure 1.1) of total motor electricity consumption (Almeida et al., 2003). In this book, electric motors, pumps and fans are researched and discussed from the energy-efficiency point of view: the energy-efficiency standards and testing methods are discussed in detail.

Roughly 30 million new electric motors are sold each year for industrial applications; some 300 million motors are in use in industry, infrastructure and large buildings. Therefore, improvements in the energy-efficiency of electric motors can reduce greenhouse gas emissions. Pumps and fans are important fluid machines, popularly used around the world, so accurately measuring the hydraulic efficiency of the pump and fan is of great importance from an economic point of view.

1.2 The history of electrical motor products

1.2.1 Electric motors

The discovery of the induction law came around 1831 from Faraday and Maxwell's formulation of the laws of electricity.

Figure 1.2 The induction motor

Faraday had a rotating homo-polar motor in 1831, which is the first electric motor in the world. The induction motor has two fathers: Galileo Ferraris (1885) and Nikola Tesla (1887). The electric motor greatly pushed the industry towards automation, which was an important development in the twentieth century. The induction motor is the main part of electric motors applied in industry (see Figure 1.2).

1.2.2 Pumps

A pump is a machine for raising, transferring or compressing fluids. The importance of the pump in the hydraulic field has been known since the Egyptians invented the shadoof in 2000 BC (World Pumps, 2009). The centrifugal pump is one of the most important types of pump and is widely used in power generation, environmental protection, the petrochemical industry and other various industrial fields. The first systematic investigations of rotodynamic pumps on a scientific basis were started in 1890 at the works of Sulzer Brothers in Switzerland. After that, other factories endeavoured to become leading manufacturers of centrifugal, helicoidal, diagonal and axial flow pumps.

In North America, the largest association of pump industry manufacturers is the Hydraulic Institute, which was founded in 1917. The European Working Group of Pump Manufacturers (Europump) was established in 1959. Europump has grown massively in stature and effectiveness since its foundation and now works in close partnership with the Hydraulic Institute in the US.

1.2.3 Fans

Fans, blowers and compressors are differentiated by the method used to move the air. Fans of all types are used in many different applications, such as ventilation and air-conditioning systems, process engineering, drying, pneumatic conveying, combustion air supply, agriculture, cooling of electronic equipment, etc. For example, the axial fan is broadly used in the ventilation field (see Figure 1.3). The first historical record concerning fans appeared in a book on mining written by Agricola in 1550. Ventilation fans first appeared about 150 years ago, initially used in mine ventilation. In 1951, the Air Movement and Control Association (AMCA) was established to improve the quality of fans. From the beginning, the AMCA implemented a Certified Rating Programme for fan performance. Certified fan performance was accepted instantly by the HVAC industry in the US and Canada. As the AMCA spread outside of North America, the AMCA certified performance of fans is also now accepted in other regions of the world, including Europe, Asia and India. The UK Fan Manufacturers Association and US ASHRAE also focus on fan development.

Figure 1.3 Axial fan

The European Energy-related Products (ErP) Directive and the forthcoming American green building codes will continuously require higher fan efficiencies, which in turn drive fan manufacturers to adopt new turbomachinery design tools and techniques.

1.3 Energy-efficiency is a top priority

1.3.1 The energy-efficiency situation of electric motors

Although electric motors can be classified into different types, the AC, three-phase induction motors are the mainstream in the global market. Three-phase induction motors constitute the large majority of motors over 0.75kW (1Hp) sold worldwide (well over 80 per cent) and are responsible for most of the motor electricity use. Therefore, the mandatory regulations and voluntary agreements have focused on the induction motor. New efficiency levels have been defined in Standard IEC 60034–30:2008, based on test methods and limits indicated under IEC 60034–2–1:2007, as follows:

- IE3 = premium efficiency;
- IE2 = high efficiency;
- IE1 = standard efficiency.

Table 1.1 **Current energy-efficiency classifications for induction motors**

Standard	Class	Power range	Efficiency interval (4 poles, 50 Hz)%
IEC	IE3 (Premium)	0.75–375kW	82.5–96
	IE2 (High)	0.75–375kW	79.6–95.1
	IE1 (Standard)	0.75–375kW	72.1–94
NEMA	NEMA Design B, General purpose electric motors	NEMA MG-1 (2006) Table 12-11 (201–500Hp)	95–95.8
	General purpose electric motors (Subtype II)	NEMA MG-1 (2006) Table 12-11 (1–200Hp)	82.5–95
	General purpose electric motors (Subtype I)	NEMA MG-1 (2006) Table 12-12 (1–200Hp)	85.5–95.8
EU Regulation	IE3	2015–2017 (7.5–375kW) 2017– (0.75–375kW)	82.5–96
	IE2	June 2011 – January 2015 (0.75–375kW) 2015–2017 (7.5–375kW), equipped with VSD 2017– (0.75–375kW), equipped with VSD	79.6–95.1

1.3.2 The energy-efficiency situation of pumps

According to a study from the American Hydraulics Institute, 20 per cent of the consumed energy has been consumed by pumps in developed countries. It has been explained that 30 per cent of this energy can be saved with good design of systems and the choice of suitable pumps. This situation has led to new research to find more-efficient systems in production and operation by producers and users of pumps (Kaya et al., 2008).

Centrifugal pumps account for 80 per cent of all pumps and it is a known fact that most centrifugal pumps have an overcapacity of 20–30 per cent. It has been calculated that the energy wasted by all the pumps operating at present in the EC is 46TWh on a yearly basis. The biggest power station within the Netherlands is the Eemscentrale in Delfzijl, with a generating capacity of 2,400MW. If the Eemscentrale operates 365 days per year continuously at full capacity, the electricity production of this station would not amount to more than 21TWh per year. So, the equivalent of more than two larger or several smaller power stations permanently operating are needed just to generate all the energy that is wasted by pumps in Europe. The pump has potential to improve running efficiency.

1.3.3 The energy-efficiency situation of fans

The energy usage by fans has been calculated to be nearly 20 per cent of the total usage of electricity – second only to pumps – so that it becomes urgent to reduce fan energy consumption as far as possible. Aerofoil bladed fans can achieve a static efficiency of up to about 85–88 per cent; by contrast, the best forward curved bladed centrifugal fan can only achieve 60 per cent. The backward bladed centrifugal fan can achieve high efficiency. According to the EU Commission, the energy consumption of fans is currently around 410TWh per annum in Europe alone, and this is expected to rise to about 660TWh by 2020. The report identified that fan efficiency can be improved by 3.5–8.3 per cent, but the system can be improved by 17.5 per cent. Energy-efficiency of fans is a function of flow, even for operation at constant speed. At very low flow, this efficiency is low but increases with increasing flow. Fan efficiency grades (FEG) were defined as areas between two adjacent efficiency limits (see Figure 1.4). The limits were shaped to follow the efficiency dependent on the fan size. The fan study proposed a minimum efficiency for each fan category as

Figure 1.4 **Fan efficiency grades (FEG) for bare shaft fans**

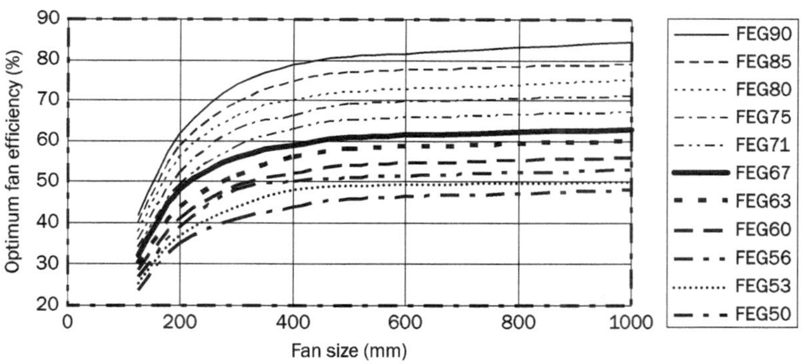

Notes

1. Fan size is the impeller diameter.
2. The optimum (peak) fan efficiency can be calculated from the total fan pressure.
3. The FEG label for a given fan size is assigned when the optimum (peak) fan efficiency is equal or lower than the efficiency at the grade upper limit and higher than the efficiency at the grade lower limit for the fan size.
4. No label is considered for the fans with the optimum (peak) efficiency below FEG53.
5. For any fan size equal or larger than 40 in., the values of the grade limits are the same as for the fan size of 40 in.
6. If this method is used for a direct driven fan without the shaft and bearings integral to the fan, the fan efficiency is the impeller efficiency.

Figure 1.5 **Axial, forward curved centrifugal, radial and mixed flow driven fan efficiency grades**

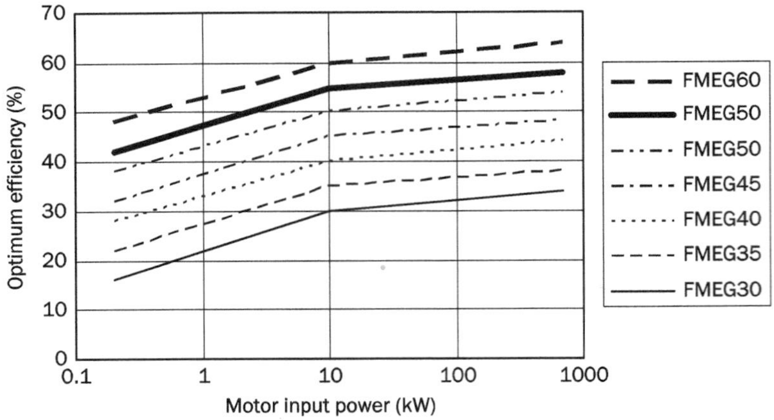

being a line from 0.125kW to 500kW, a minimum efficiency line (MEL), later called a fan motor efficiency grade (FMEG) by the fan industry (see Figure 1.5) (AMCA, 2010).

1.4 Conclusion

Energy-efficiency improvement plays an important role in a sustainable energy future, including energy security, economic development and environmental protection. Both developing economies, such as China and India, and developed countries, such as the USA and Japan, face acute energy shortages, which is an obstacle to sustainable economic development. Electricity consumption for electrical motor products occupies a large proportion of global electricity consumption. Although the energy-efficiency of induction motors, pumps and fans has attracted much more attention by independent organizations, there is still room for improvement from the current situation with electrical motor products.

2

Classification: electric motors, pumps, fans

Abstract: This chapter presents definitions of the electric motor, pump and fan. The classification of the electric motor, pump and fan is discussed respectively. Different types of induction motor, pump and fan also are discussed. The working principles of the electric motor, pump and fan are introduced.

Key words: electrical motor products, definition, classification, working principle.

2.1 Definition of electric motors

An electric motor is a machine used to transfer electrical energy into mechanical energy. An electric motor requires an interaction between the rotating parts and the stationary parts. Interaction between the magnetic fields of these electromagnets generates twist force, or torque; as a result, the motor rotates in the direction of the resultant torque (Holopainena et al., 2005).

Motors can operate on either a single phase or three phases, direct or alternating current, at several voltages. The voltage and the type of current used depend on the size of the motor and the electrical service available. The phase refers to the number of cycles of alternating current.

Figure 2.1 Complete cycles of three-phase alternating current

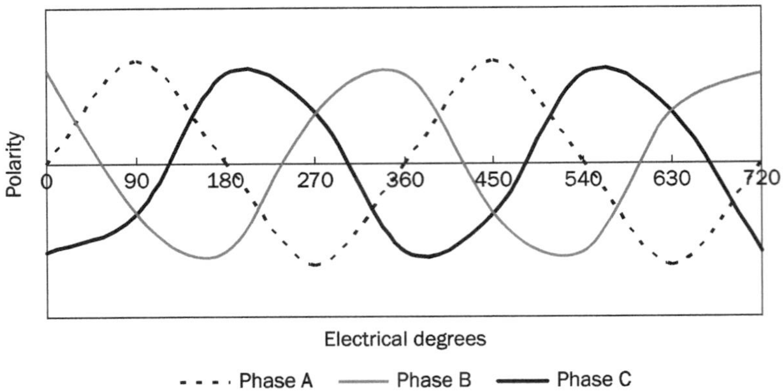

In three-phase alternating current service, three single phases are combined so the peak voltages are an equal distance apart (see Figure 2.1). Three-phase current is recommended over single phase for larger loads and is required by most electrical service companies for motors over 5–10 horsepower.

Synchronous machines are principally used as alternating current (AC) generators. Synchronous machines can also be used as motors, but they are usually built in very large sizes. The synchronous motor operates at a precise synchronous speed, and hence is a constant-speed motor. Unlike the induction motor, whose operation always involves a lagging power factor, the synchronous motor possesses a variable-power-factor characteristic and hence is suitable for power-factor correction applications. Three-phase induction motors are simple, rugged, low-cost and easy to maintain. The common types of electric motors are shown in Figure 2.2.

Figure 2.2 Common types of electric motor

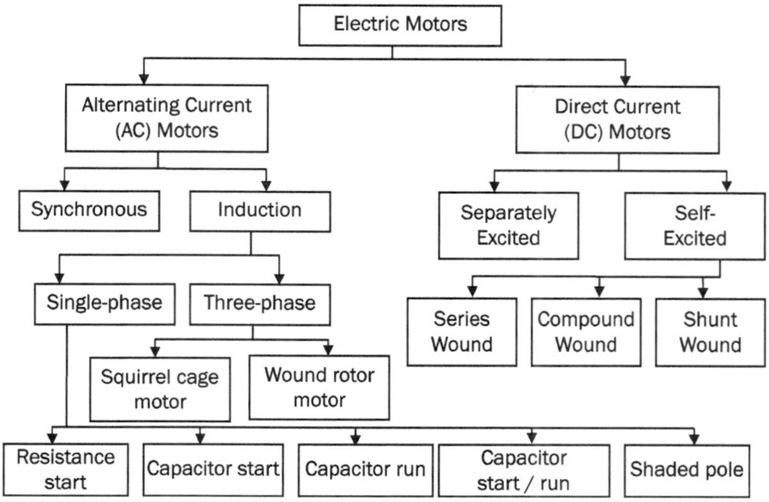

2.2 Working principles of the induction motor

Operation of three-phase induction motors is based upon the application of Faraday's Law and the Lorentz Force on a conductor. In an AC induction motor, one set of electromagnets is formed in the stator because of the AC supply connected to the stator windings. The alternating nature of the supply voltage induces an electromagnetic force (EMF) in the rotor (just like the voltage is induced in the transformer secondary) as per Lenz's law, thus generating another set of electromagnets; hence the name – induction motor. The equivalent circuit of the induction motor is shown in Figure 2.3 (Sujitjorn and Areerak, 2004). Referring to Figure 2.3, s is slip, while R_i and L_i are motor parameters. These parameters are assumed constant conventionally and obtained from the no-load and the blocked-rotor tests.

Figure 2.3 Equivalent circuit of an induction motor

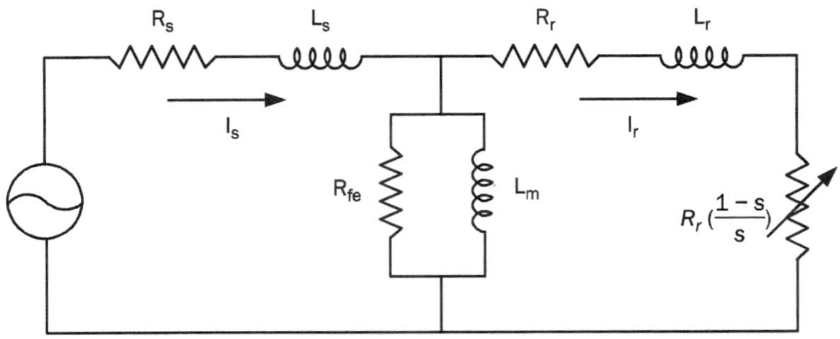

Realistically, only the core loss term represented by R_c and L_m may assume constant values.

Like most motors, an AC induction motor has a fixed outer portion, called the stator, and a rotor that spins inside with a carefully engineered air gap between the two.

Virtually all electrical motors use magnetic field rotation to spin their rotors. A three-phase AC induction motor is the type where the rotating magnetic field is created naturally in the stator because of the nature of the supply. DC motors depend either on mechanical or electronic commutation to create rotating magnetic fields.

2.3 Types of induction motor

2.3.1 Single-phase induction motors

A single-phase AC induction motor depends on extra electrical components to produce this rotating magnetic field. Two sets of electromagnets are formed inside any motor. As illustrated in Figure 2.2, the single-phase induction motor is classified into resistance start, capacitor start, capacitor run, capacitor start/run and shaded pole. These different types of single-phase induction motor are described below.

Resistance-start motors, also called split-phase motors (see Figure 2.4), use both an auxiliary and a main winding. The auxiliary winding is displaced 90 electrical degrees from the main winding. The main winding has many turns of large diameter wire wound in the bottom of the stator slots to get high reactance, which is an inductive circuit. The auxiliary winding has few turns of small diameter wire wound to get high resistance, which is a resistive circuit. Therefore, the current in the auxiliary winding leads the current in the main winding, causing a rotating field. During start-up, both windings are connected to the line. As the motor comes up to a certain speed (for example, at about 70 per cent of full-load speed), a centrifugal switch or an electronic switch disconnects the auxiliary winding. Resistance start motors are considered low or moderate starting torque motors and are limited to about 1/3hp.

Capacitor-start motors also belong to split-phase motors, as illustrated in Figure 2.5. The main difference compared with the resistance start motor is that a capacitor is placed in series with the auxiliary winding; bigger capacitor value will produce bigger start torque. The function of the auxiliary winding is to make the motor start. When the rotation speed of the motor reaches certain value, the centrifugal switch cuts

Figure 2.4 **Resistance-start AC induction motor**

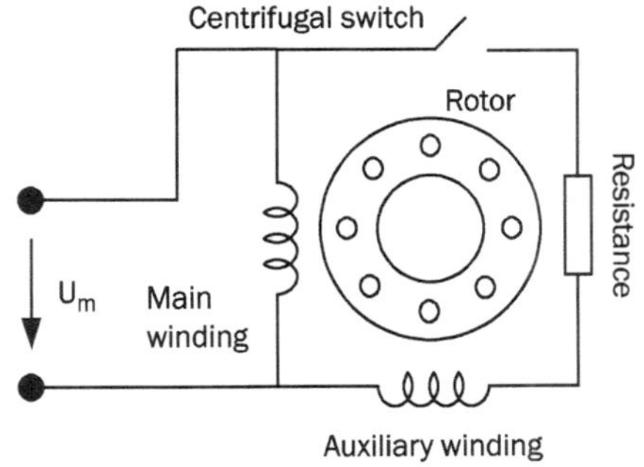

Figure 2.5 Capacitor-start AC induction motor

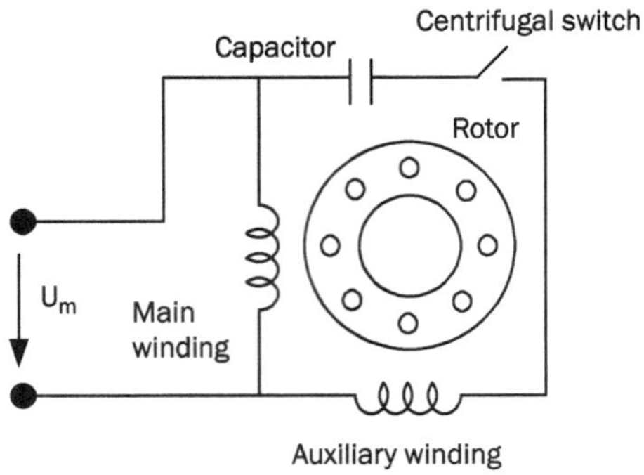

off the power supplied to the auxiliary winding, then the main winding works independently. This type of motor produces greater locked rotor and accelerating torque per ampere than does the resistance-start motor. Sizes range from fractional to 10hp at 900 to 3,600rpm. The design is a heavy-duty unit, which has approximately 300 per cent (of full load) starting torque. Common applications include compressors, pumps, conveyors and other 'hard-to-start' applications.

Capacitor-run motors are the ones which have the earliest and widest applications among the single-phase motors (see Figure 2.6). Capacitor-run motors use a capacitor connected

Figure 2.6 The capacitor-run AC induction motor

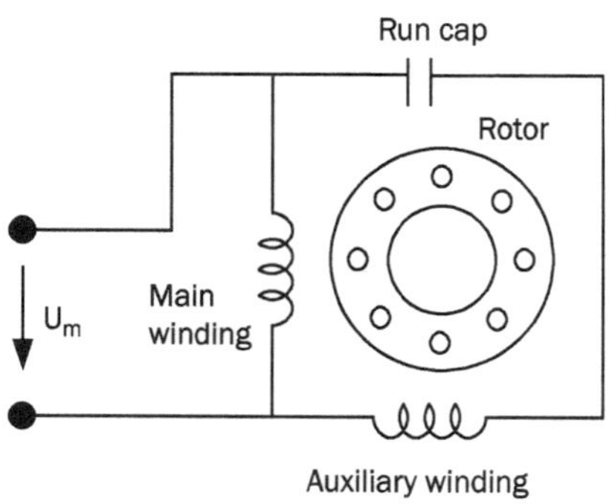

Figure 2.7 **The capacitor in the capacitor-run AC induction motor**

in series with the auxiliary windings (see Figure 2.7). The fundamental function of such a capacitor is to create a phase shift between currents in the main and auxiliary windings. This leading phase shift is necessary for enabling the motor to develop a sufficiently high starting electromagnetic torque and operate as a balanced two-phase machine.

The capacitor start/run AC induction motor can be designed for lower full-load current and higher efficiency (see Figure 2.8). This motor is costly due to start and run capacitors and centrifugal switch. It is able to handle applications too demanding for any other kind of single-phase motor. These include wood-working machinery, air compressors, high-pressure water pumps, vacuum pumps and other high-torque applications requiring 1–10hp. This type of design has lower full-load amps as a result of the run

Figure 2.8 Capacitor start/run AC induction motor

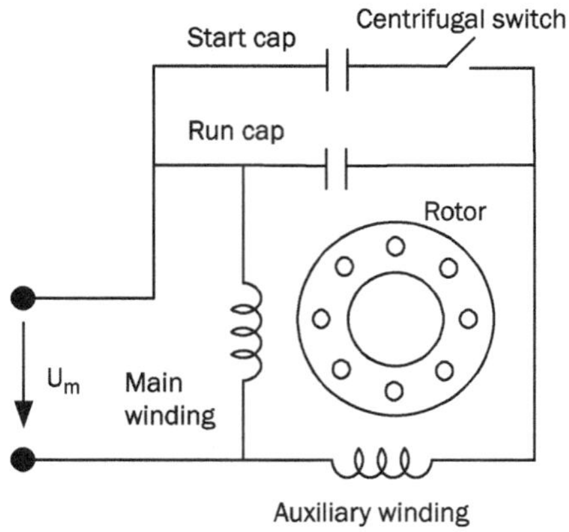

capacitor and is consequently used on mostly higher horsepower single-phase motors.

Shaded pole motors have a continuous copper loop wound around a small portion of each pole (see Figure 2.9). The loop causes the magnetic field through the ringed portion to lag behind the field in the unringed portion. This produces a slightly rotating field in each pole face sufficient to cause the rotor to revolve. As the rotor accelerates, its torque increases and its rated speed is obtained. Shaded pole motors have low starting torque and efficiency ratings, which are available only in fractional and subfractional horsepower sizes. Slip is

Figure 2.9 Shaded pole AC induction motor

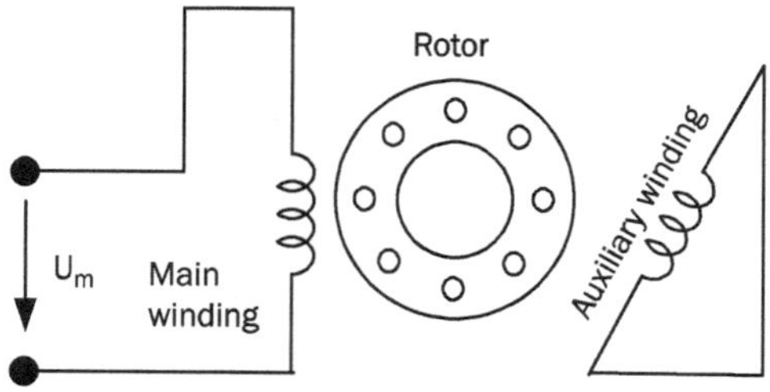

Figure 2.10 **Original photo of shaded pole AC induction motor**

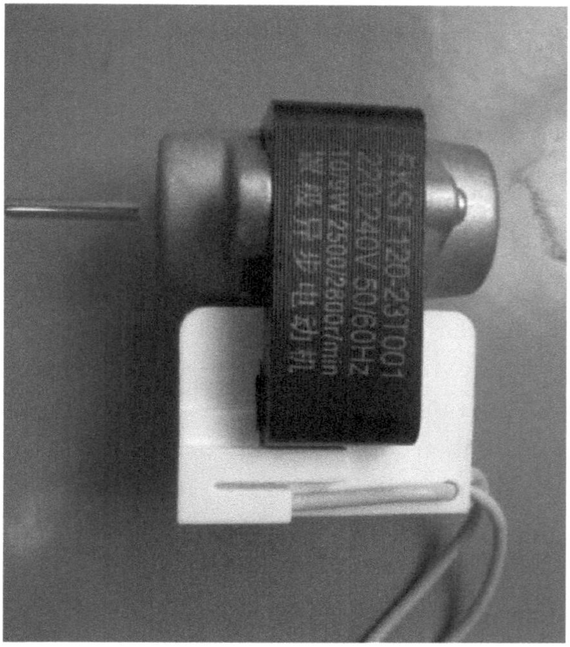

about 10 per cent or more at rated load. Shaded pole motors have the advantages of simple structure, high reliability and low cost, which can be applied in the field of low power and low start torque. The original photo of a shaded pole AC induction motor is shown in Figure 2.10.

2.3.2 Three-phase induction motors

The type of winding gives rise to two main types of motors: squirrel cage induction motors and wound rotor induction motors. A three-phase induction motor has two main parts:

- a stator – consisting of a steel frame that supports a hollow, cylindrical core of stacked laminations; slots on the internal circumference of the stator house the stator winding;

- a rotor – also composed of punched laminations, with rotor slots for the rotor winding.

There are two types of rotor windings: squirrel-cage windings and conventional three-phase windings.

Squirrel-cage windings produce a squirrel-cage induction motor (most common). Squirrel-cage induction motors are most common and are either three phases or single phase. Almost 90 per cent of the three-phase AC induction motors are of this type. Motors of this type, rated 1 horsepower or larger, cost less and can start at heavier loads than their single-phase counterparts. Most induction motors contain a rotor in which the conductors, made of either aluminium or copper, are arranged in a cylindrical format resembling a 'squirrel cage'. Squirrel-cage induction motors have no external electrical connections to the rotor, which is made of solid, un-insulated aluminium or copper bars short-circuited at both ends of the rotor with solid rings of the same metal. The rotor and stator are connected by a magnetic field that crosses the air gap. This simple construction results in relatively low maintenance requirements.

The relationship between torque and speed in squirrel-cage motors is largely dependent on rotor resistance. As the rotor resistance decreases, the performance speed improves and the starting torque decreases. The smaller the slip for a given load, the higher the efficiency because induced currents and their associated rotor losses are also smaller. Three-phase squirrel-cage induction motors dominate application above 1hp. Single-phase squirrel-cage induction motors are more common in sizes below 1hp and large home appliances. Single-phase motors are larger and more expensive, with a lower efficiency than three-phase motors that have the same power and speed ratings. Additionally, three-phase induction motors are more reliable since they do not need special starting equipment.

Conventional three-phase windings made of insulated wire produce a wound-rotor induction motor (special characteristics) (see Figure 2.11).

The wound-rotor motor is a variation of the squirrel-cage induction motor. While the stator is equivalent to that of the squirrel-cage motor, it has a set of windings on the rotor which are not short-circuited, but are terminated to a set of slip rings. These are helpful in adding external resistors and contactors. The slip necessary to generate the maximum torque (pull-out torque) at lower speed is directly proportional to the rotor resistance. As the motor accelerates, the value of the resistance can be reduced, altering the motor characteristic to suit the load requirement. Once the motor reaches the

Figure 2.11 A wound-rotor induction motor

base speed, external resistors are removed from the rotor. This means that now the motor is working as the standard induction motor. This motor type is ideal for very high inertia loads, where it is required to generate the pull-out torque at almost zero speed and accelerate to full speed in the minimum time with minimum current draw.

2.4 Definition of pumps

The pump has two basic types – rotodynamic and positive displacement pumps. The pump type tree is shown in Figure 2.12. Rotodynamic pumps are generally divided into three classes: radial flow, axial flow and mixed flow. The relationship of pump developed head with the pump discharge flow at constant speed in general is called the pump performance characteristic. Head is highlighted here because pump people always talk about 'head' but mean 'differential head'.

A pump senses the liquid head at the inlet connection and adds the developed head to it. The pressure in the pump

Figure 2.12 Pump type tree

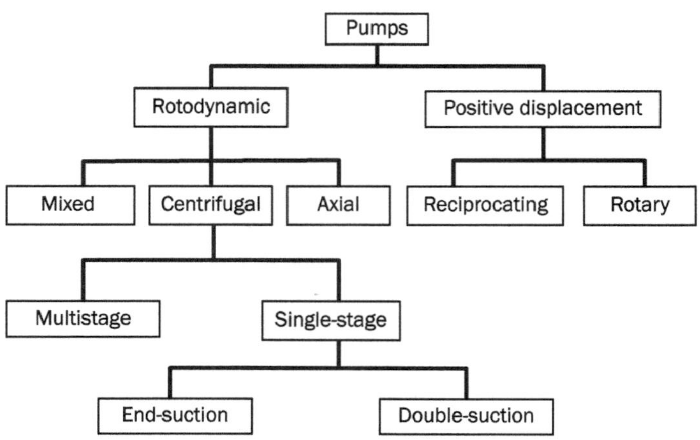

outlet is a combination of the inlet conditions and the developed head. The centrifugal pump selection guide is shown in Figure 2.13.

If a rotodynamic pump is operated with very low differential head, it is likely the flow will increase dramatically and so will the absorbed power. Pump drivers may need to be oversized to accommodate 'runout', which may occur during start-up.

Positive displacement pumps have very different 'head-flow' characteristics. A positive displacement pump delivers at a fixed speed a nearly constant flow rate independently of the back pressure. The positive displacement pumps (e.g. piston pumps) essentially deliver the same volume V_{stroke} at each stroke independently of flow velocity or rotor speed n. The flow rate then becomes $Q = n \times V_{stroke}$; the pressure rise results solely from the imposed back pressure.

Figure 2.13 **Centrifugal pump selection guide**

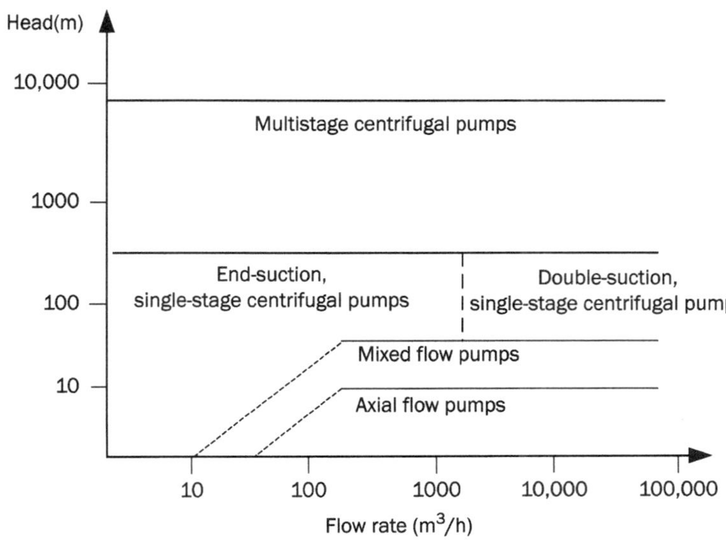

2.5 Working principles of pumps

The flow rate of a pump will be measured and guaranteed when the pump is tested. Rated flow is associated with rated differential head for rotodynamic pumps and rated outlet pressure for positive displacement pumps. For a fixed-speed rotodynamic pump, the pump is capable of operating at a range of flows with corresponding differential heads. The ideal pump selection would have the pump operating at its best efficiency point (BEP).

The flow rate of pump can be adjusted easily by changing the differential head. Throttling the pump outlet is common practice, although it is very inefficient. The design of an efficient pumping system depends on relationships between fluid flow rate, piping layout, control methodology and pump selection. The complete definition of the pump performance includes the efficiency, the NPSH and the power requirements over the flow range. The following relationships are useful in providing the performance of centrifugal pumps:

- flow rate is directly proportional to the rotation speed S;
- differential head is directly proportional to the rotation speed S^2;
- power required is directly proportional to the rotation speed S^3.

2.6 Types of rotodynamic pump

2.6.1 Radial-flow/centrifugal pump

The radial-flow pump, also named the centrifugal pump, is the most popular style of rotodynamic pump (see Figure 2.14).

Figure 2.14 A centrifugal pump

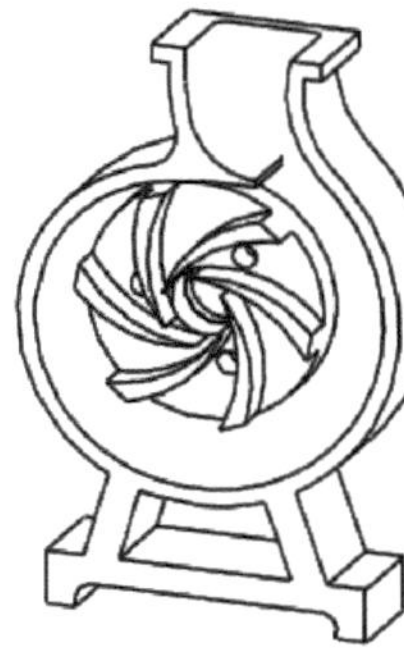

It directs the liquid flow to be radial (hence radial-flow pump), perpendicular to the shaft axis, and relies on centrifugal force to create kinetic energy.

Centrifugal pumps are turbomachines applied for transporting liquids by raising a specified volume flow to a specified pressure level. The energy transfer in turbomachines is invariably based on hydrodynamic processes for which characteristically all pressure and energy differences are proportional to the square of the circumferential rotor speed (Gulich, 2010).

The single-stage centrifugal pump consists of four main components: the inlet with the inducer or the inlet without the inducer, the impeller attached to a rotating shaft, the vaneless and vaned diffuser, and the volute enclosing the impeller. Centrifugal pumps can be manufactured as single-stage or multi-stage. For the single-stage centrifugal pumps, only one impeller is mounted on the rotor shaft. For the multi-stage centrifugal pumps, several impellers are mounted on the rotor shaft and a return channel is used between the two single-stage centrifugal pumps.

Centrifugal pumps deal with high flow rates, provide smooth, non-pulsating delivery and regulate the flow rate over a wide range without destroying the pump. Centrifugal pumps have few moving components and the wear caused by normal operation is minimal. They are also compact and easily disassembled for maintenance.

2.6.2 Axial flow/propeller pump

In 1785, John Skeys patented a pump of novel construction, which was the prototype of the axial flow pump. The English physicist and engineer, James Thomson, initiated the idea of using guide vanes to enhance the performance of pumps. The English scientist Osborne Reynolds designed the first pump with adjustable inlet guide vanes in 1875 (Kaya, 2003). The first systematic investigations of rotodynamic pumps on a scientific basis were commenced in the 1890s at the works of Sulzer Brothers in Switzerland. This lead was followed by other factories that endeavoured to become leading manufacturers of centrifugal pump design and, after this, the design of helicoidal, diagonal and axial flow pumps was developed rapidly.

As a general rule, axial flow pumps are usually selected for pumping large volumes of water against relatively low heads (1–15m). The total head increase for axial flow pumps is small compared with that of centrifugal pumps.

2.6.3 Mixed-flow/screw pump

Mixed-flow (MF) pumps are hybrids, in that they use both mechanisms, which is why they are sometimes referred to as diagonal pumps. A benefit of the hybrid nature of MF pumps is that they can be used with both volute (radial) housings and straight-through tubular (axial) housings. In conjunction with dry or wet sump intakes, they offer great freedom in installation design, which is often a critical cost factor.

MF pumps generally exhibit a steep rise of pressure from their best efficiency point (BEP) to zero flow and can therefore have relatively high power at zero flow, compared with centrifugal pumps. Also, the steep head characteristic allows MF pumps to operate well in a parallel arrangement. Their maximum power point is close to their BEP. Another important advantage of MF pumps is that they can be designed to be reversible, so that the same unit can be used both as pump and as turbine, depending on the direction of flow. Because of their flexibility and long-term reliability, MF pumps are widely used for applications requiring high-volume flow against moderate heads.

2.7 Definition of fans

A fan is one kind of turbomachinery. Fan means a rotary-bladed machine that is used to maintain a continuous flow of gas, typically air. The main types of fan are characterized by the path of the airflow through the fan. According to the air flow direction in the fans, three fan types are classified: centrifugal, axial fan and mixed-flow fan. Centrifugal fans can be applied in the fields which need low flow rate and high pressure, while axial fans are usually used in the fields which require high flow rate and low pressure.

2.8 Working principle of fans

The purpose of a fan is to move air continuously against a resistance. A fan must increase the pressure of the air by the amount necessary to overcome this resistance. Fans use rotation blades to push the flow of gas. The terminology and definition of category standard ISO/DIS13349:2008 refers to more than 20 fan types. These can be simplified to eight types: axial, radial, forward curved centrifugal, backward bladed centrifugal (with and without housing), mixed flow and tangential.

2.9 Types of fan

2.9.1 Centrifugal fans

Centrifugal fans consist of an impeller in a casing having a spirally shaped contour. The air enters the impeller in an axial direction and is discharged at the impeller outer periphery. The air flow moves along the centrifugal direction (or radial direction). Sometimes centrifugal fans are called radial flow type fans. Centrifugal fans are capable of generating relatively high pressures. They are suitable for high pressure applications as compared with axial flow fans. Generally centrifugal fans have three types of blade: forward blade, backward blade and radial blade. The characteristic curve of three kinds of centrifugal fan is shown in Figure 2.15.

Figure 2.15 Fan static pressure and air volume for different centrifugal fans

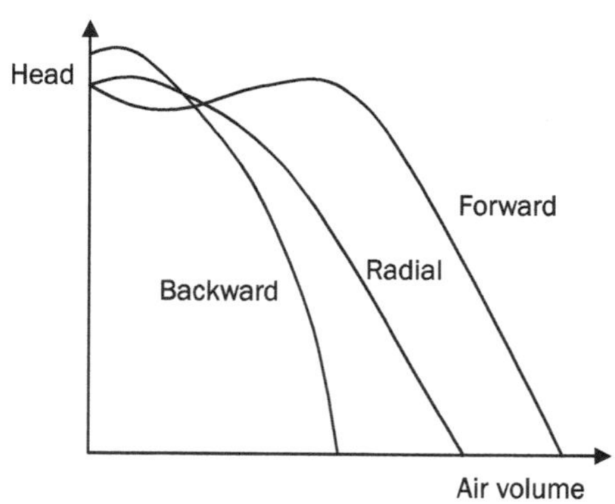

They are frequently used in 'dirty' airstreams (high moisture and particulate content), in material handling applications and in systems at higher temperatures. They can be small, with sub 0.125kW motors, or large, with hundreds of kW motors driving fans. The characteristic of different centrifugal fans are shown in Table 2.1 (United Nations Environment Program, 2006; Eskom, 2010).

Table 2.1 **Characteristics of different centrifugal fans**

Type of fan	Applications	Advantages	Disadvantages
Radial 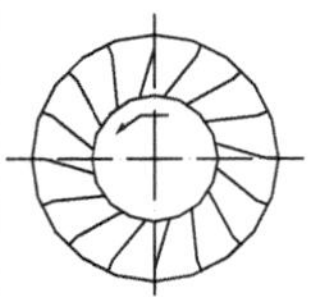	(1) Industrial/ mining/ petrochemical (2) Dust extraction applications (3) High-pressure applications	(1) Suitable for high static pressures and high temperatures (2) Simple design allows custom-build units for special applications (3) Can operate at low air flows without vibration problems (4) High durability (5) Have large running clearances, which is useful for airborne solids (dust, wood chips and metal scraps) handling services	Only suitable for low–medium airflow rates
Forward curved 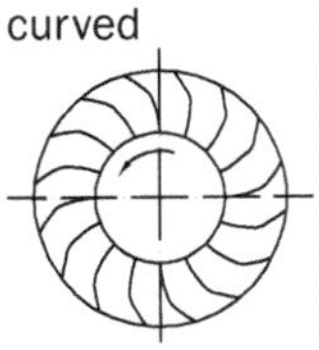	(1) HVAC/ industrial (2) Clean air supply & and extraction	(1) Can move large air volumes against relatively low pressure (2) Relative small size	(1) Only suitable for clean service applications but not for high pressure and harsh services

(continued)

Table 2.1 Characteristics of different centrifugal fans (*cont'd*)

Type of fan	Applications	Advantages	Disadvantages
		(3) Low noise level (due to low speed) and well suited for residential heating, ventilation, and air-conditioning (HVAC) applications	(2) Fan output is difficult to adjust accurately (3) Driver must be selected carefully to avoid motor overload because power curve increases steadily with airflow (4) Relatively low energy-efficiency (55–65%)
Backward inclined 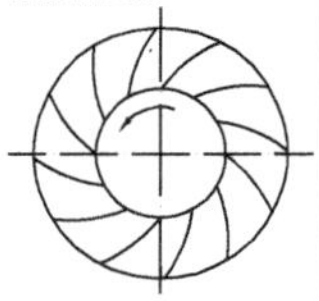	(1) HVAC/ industrial (2) Clean air supply and extraction	(1) Can operate with changing static pressure (as this does not overload the motor) (2) Suitable when system behaviour at high air flow is uncertain (3) Suitable for forced-draft services (4) Flat-bladed fans are more robust (5) Curved-bladed fans are more efficient (exceeding 85%) (6) Thin air-foil bladed fans are most efficient	(1) Not suitable for dirty air streams (as fan shape promotes accumulation of dust) (2) Airfoil bladed fans are less stable because of staff as they rely on the lift created by each blade (3) Thin airfoil bladed fans subject to erosion
Aerofoil blades	(1) Industrial/ mining/ petrochemical (2) Clean air supply and extraction	(1) High efficiencies (85–90%) (2) High volume and medium to high pressure (3) Non-overloading characteristics	(1) Extremely high labour input to manufacture

Figure 2.16 Centrifugal fan with scroll enclosure

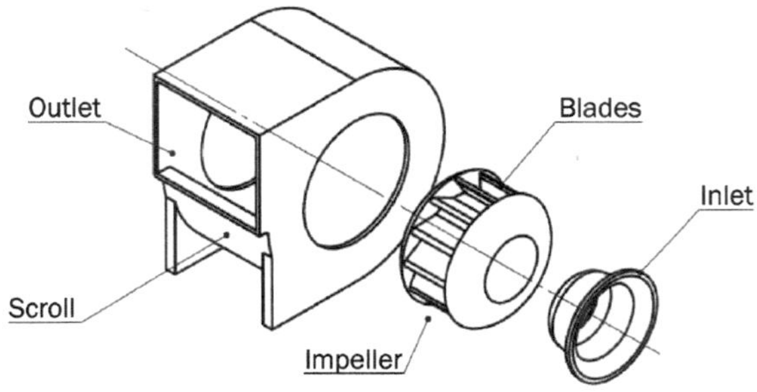

Over the last 20 years centrifugal fans with backward curved blades have become more common within air handling. They are normally larger for the same volume and pressure characteristics and are a little more complicated to construct and so more expensive. Despite the cost disadvantage, they have become more popular in air handling units because their efficiency is higher than forward curved centrifugal fans. Centrifugal fans with forward curved blades always require a scroll housing (see Figure 2.16). Their characteristic features are a low noise level, relatively small air flow with a high increase in static pressure and space-saving application.

2.9.2 Axial fans

The air stream in the axial fan moves along the axial direction of the fan. The mechanical energy of the driver is converted into spiral airflow at the fan discharge. The air is pressurized by the aerodynamic lift generated by the fan blades, much like a propeller and an aeroplane wing (see Figure 2.17). Although they can sometimes be used interchangeably with centrifugal fans, axial fans are commonly used in 'clean air', low-pressure, high-volume applications. Axial fans have less rotating mass and are more compact than centrifugal fans of comparable capacity. Additionally, axial fans tend to have higher rotational

Figure 2.17 Vane-axial fan

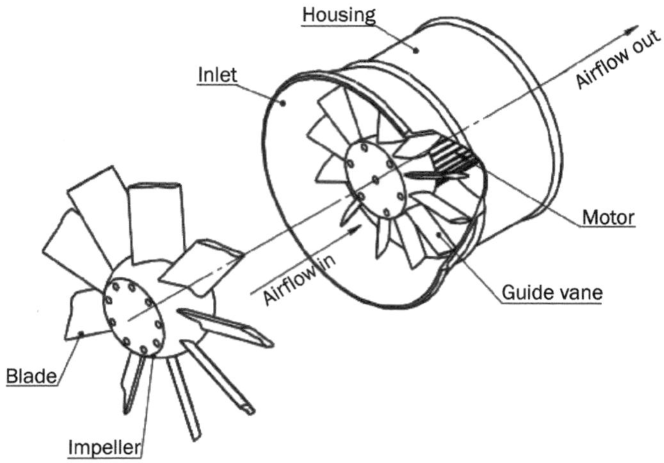

speeds and are somewhat noisier than in-line centrifugal fans of the same capacity; however, this noise tends to be dominated by high frequencies, which tend to be easier to attenuate.

Axial fans provide airflows over an extremely large range of flow rates and pressure rise values. Their distinguishing feature is that the pressure rise will be relatively small and the flow rate relatively large in comparison with centrifugal fans. The characteristic of different axial fans is illustrated in Table 2.2 (United Nations Environment Program, 2006; US DOE, 2003).

By comparison, a centrifugal or radial flow fan moves air perpendicular to the axis of rotation. Axial flow fans are better suited for low-resistance, high-flow applications, whereas centrifugal flow fans apply to high-pressure resistance, low-flow conditions.

2.9.3 Mixed-flow fans

In mixed-flow fans the air flow moves between the centrifugal type and the axial type. Mixed-flow fans are sometimes

Table 2.2 **Characteristics of different axial fans**

Types	Advantages	Disadvantages	Efficiency range and applications
Propeller	(1) Generate high airflow rates at low pressures (2) Not combined with extensive ductwork (because they generate little pressure) (3) Inexpensive because of their simple construction (4) Achieve maximum efficiency, near-free delivery, and are often used in rooftop ventilation applications (5) Can generate flow in reverse direction, which is helpful in ventilation applications	(1) Relatively low energy-efficiency (2) Comparatively noisy	40–50% Often used in rooftop ventilation applications
Vane-axial	(1) Suited for medium- to high-pressure applications (up to 500 mm WC), such as induced draft service for a boiler exhaust (2) Can quickly accelerate to rated speed (because of a low rotating mass) (3) Suited for direct connection to motor shafts	(1) Relatively expensive compared with propeller fans	78–85% Used in a boiler exhaust Useful in emergency ventilation applications, and generate flow in reverse directions, which is useful in many ventilation applications

(continued)

Table 2.2 Characteristics of different axial fans (*cont'd*)

Types	Advantages	Disadvantages	Efficiency range and applications
	(4) Most energy-efficient (up to 85% if equipped with airfoil fans and small clearances)		
Tube-axial	(1) Higher pressures and better operating efficiencies than propeller fans (2) Suited for medium-pressure, high airflow rate (3) Can quickly accelerate to rated speed (because of their low rotating mass) and generate flow in reverse direction, which is useful in many ventilation applications (4) Create sufficient pressure to overcome duct losses and are relatively space efficient, which is useful for exhaust applications	(1) Relatively expensive (2) Moderate airflow noise (3) Relatively low energy-efficiency	67–72% Many ventilation applications, well suited for ducted HVAC installations

called lean-type fan. A mixed-flow fan is an axial flow impeller modified to generate a radial component of flow, which is added to the spiral flow, achieving a capacity for high volume and high pressure.

2.10 Conclusion

In this chapter the types of induction motor, pump and fan have been described. The advantages and disadvantages among the different types of induction motor, pump and fan have been analysed, and selection guides have been provided for using them.

3

Energy-efficiency technical measures system for electrical motor products

Abstract: The framework of the energy-efficiency technical measures is analysed in this chapter. Regulations, agreements, standards, labels, testing methods and incentive programmes for the energy-efficiency of electrical motor products are put forward.

Key words: energy-efficiency, technical measures, regulations, standards, testing methods.

3.1 The framework of energy-efficiency technical measures for electrical motor products

In the definition of the WTO Agreement, the Technical Barriers to Trade (the 'TBT' Agreement) includes the product standards, technical regulations and the procedures (such as testing and certification). The energy-efficiency technical measures for electrical motor products consist of regulations/agreements, standards, testing methods and labels. The regulations/agreements are located at the top level among these measures. Efficiency testing methods are the foundation for regulations/agreements, efficiency standards, energy labels and other related programmes, which provide the manufacturers, regulatory authorities and consumers a way of consistently evaluating the status of energy use with different models.

The establishment of a certain efficiency standard is based on the efficiency survey and accumulated test data of a certain product. Many appliances have been tested with one of the test procedures when they are sold in the market. The test data can also be used to accelerate the time of setting standards. The third step is to develop an energy label. Labels are useful to educate consumers; however, they are not absolutely necessary for standards. Energy labels could contribute to further potential savings from the standards. The last step is to develop some incentive programmes as an option for standards and labels. The hierarchy of energy-efficiency technical measures is shown in Figure 3.1 (Yanti and Mahlia, 2009). In this chapter, the energy-efficiency technical measures for electric motors are described in detailed.

Figure 3.1 The hierarchy of energy-efficiency technical measures

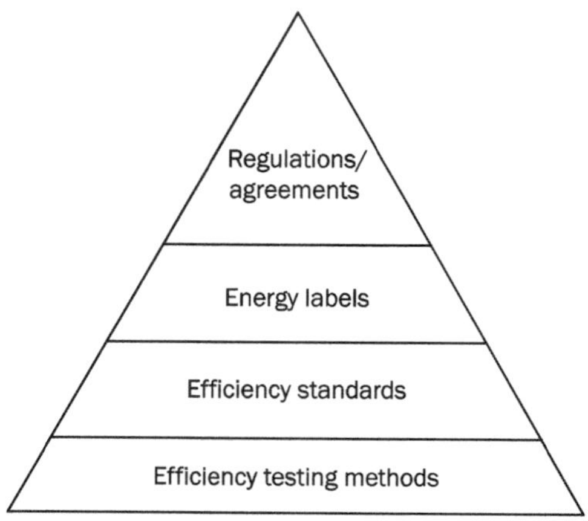

3.2 Regulations and agreements

3.2.1 Mandatory regulations for electric motors

The energy policy act (EPAct)

The Energy Policy Act, which was signed into law in 1992, came into effect on 24 October 1997. This law, for the first time, mandates energy-efficiency standards for all general-purpose motors, from 1 to 200 horsepower, which are manufactured for sale in the United States. The EPAct defines 'electric motor' to include: 'general purpose, T-frame, single-speed, foot-mounting, polyphase squirrel cage induction motor of the National Electrical Manufacturers Association (NEMA) Design A and B, continuous-rated, operating on 230/460 volts and constant 60 hertz line power, as defined in NEMA Standards Publication MG1 – 1987.' In addition, the EPAct also establishes new testing procedures and labelling requirements for electric motors. The US Department of Energy (DOE) is responsible for establishing the rules to implement and enforce the EPAct.

EU ecodesign directive

Directive 2005/32/EC established a framework for the setting of ecodesign requirements for energy-using products. Directive 2009/125/EC of the European parliament established a framework for the setting of ecodesign requirements for energy-related products, which is the further amendment of Directive 2005/32/EC. The implementation of the ErP Directive is an ongoing process. The implementing measures on electric motor and circulators have been published, while the implementing measure for ventilation fans was published in October 2010.

The European Commission has foreseen three steps for the introduction of minimum energy performance standards for electric motors in Europe based on the classification scheme defined by the IEC. European Commission regulation (EC) No. 640/2009 is the mandatory measure on electric motor minimum efficiency. The measure is taken in three phases (European Commission, 2009):

- Phase 1: Motors in the power range 0.75–375kW manufactured in or imported into the EU after 16 June 2011 must meet or exceed the 'IE2' efficiency level.
- Phase 2: Motors in the power range 7.5–375kW manufactured in or imported into the EU after 1 January 2015 must meet or exceed the 'IE3' efficiency level or meet or exceed the 'IE2' efficiency level when equipped with a variable speed drive.
- Phase 3: Each motor in the range 0.75–375kW manufactured in or imported into the EU after 1 January 2017 must meet or exceed the 'IE3' efficiency level or meet the 'IE2' efficiency level when equipped with a variable speed drive.

European Commission regulation (EC) No. 641/2009 sets up the minimum energy-efficiency requirements for glandless standalone circulators and glandless circulators integrated in products. The mandatory requirements are in two phases:

- Phase 1: From 1 January 2013, glandless standalone circulators, with the exception of those specifically designed for primary circuits of thermal solar systems and of heat pumps, shall have an energy-efficiency index (EEI) of not more than 0.27.
- Phase 2: From 1 August 2015, glandless standalone circulators and glandless circulators integrated in products

shall have an energy-efficiency index (EEI) of not more than 0.23.

3.2.2 *Voluntary agreements*

CEMEP

The EU-CEMEP agreement provides for the rating and identification of motor efficiency. The EFF-logo is a registered trademark by Gimélec (French motors and drive association) on behalf of CEMEP. Two efficiency indices, high and low, are stipulated for motors of different specifications. Motors with an efficiency value lower than the low index are classified as EFF3; those with an efficiency value between the low and the high index are EFF2; those above the high index are classified as EFF1. The loss of EFF1 motors is 40 per cent lower than that of EFF2 motors, and they are designed for more than 6000h of annual runtime. The loss of EFF2 motors is 20 per cent lower than that of EFF3 motors, and they are designed for more than 2,000h of annual runtime. The efficiency of EFF1 motors, compared with EFF2 motors, has improved 1–5 per cent according to different powers. EFF3 motors are generally known as low-efficiency motors, EFF2 motors as improved-efficiency motors and EFF1 motors as high-efficiency motors. The agreement also requires manufacturers to identify efficiency grade on their product nameplates and sample data sheets to facilitate selection and identification for users. CEMEP recommends using the new efficiency classes of IEC 60034-30 (IE-code), which are based on new test methods explained in IEC 60034-2-1:2007 from now.

NEMA premium

A premium class is established by the National Electrical Manufacturers Association (NEMA) with rated and minimum efficiency levels for closed and open motors from two to six poles, and includes motors up to 500hp.

The member companies of the NEMA Motor and Generator Section established a NEMA Premium® energy-efficiency motors programme to provide highly energy-efficient products that meet the needs and applications of users and original equipment manufacturers (OEMs) based on a consensus definition of 'premium efficiency' and use of the NEMA Premium® logo for premium products.

NEMA Premium® labelled electric motors will assist purchasers to optimize motor systems efficiency, reduce electrical power consumption and costs, and improve system reliability. The market scale of NEMA Premium® motor is rising in the USA, shown as Figure 3.2. It is estimated that the NEMA Premium® efficiency motor programme would save 5,800 gigawatts of electricity. This translates to preventing nearly 80 million metric tons of carbon dioxide going into the atmosphere over the next ten years – equivalent to keeping 16 million cars off the road.

Figure 3.2 **United States electric motor market distribution by motor efficiency standards**

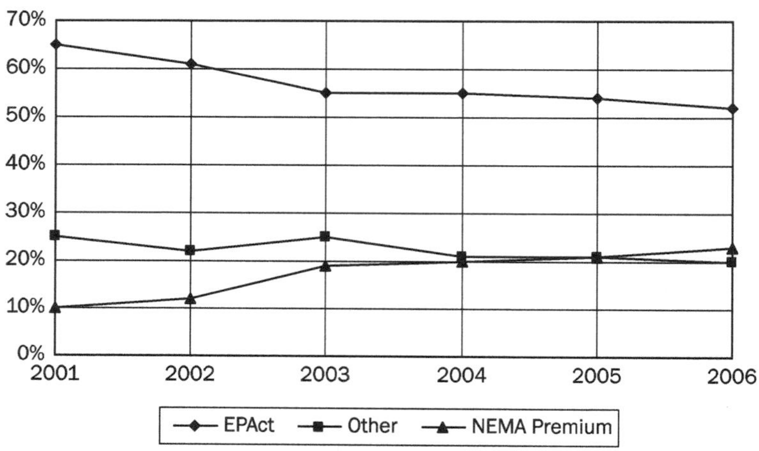

3.3 Energy-efficiency standards

Energy-efficiency standards are a set of procedures and regulations that prescribe the energy performance of manufactured products, sometimes prohibiting the sale of products that are less efficient than a minimum level. There are three types of energy-efficiency standards (Wiel and McMahon, 2003):

1. prescriptive standards – requiring that a particular feature or device be installed (e.g. insulation) or not installed (e.g. pilot lights) in all new products;
2. minimum energy performance standards (MEPS) – prescribing minimum efficiencies (or maximum energy consumption – usually as a function of size or capacity) that manufacturers must achieve in each and every product, specifying the energy performance but not the technology or design details of the product;
3. class-average standards – specifying the average efficiency of a manufactured product, allowing each manufacturer to select the level of efficiency for each model so that the overall average is achieved.

Standards are mandatory programmes (regulations) stipulating the minimum efficiency levels or maximum energy-use levels acceptable for products sold in a particular country or region. They are often called minimum efficiency standards or minimum energy performance standards (MEPS). Though regulatory standards are most often applied to all products on a given market, the product coverage could conceivably be a given percentage of products on the market, a market-wide average or a manufacturer-based average.

3.4 Energy-efficiency labels

A label is a mandatory or voluntary sticker that is affixed to products or their packaging containing information on the energy-efficiency or energy consumption of the product (Mahlia et al., 2002). Usually labels enable the comparison of energy-efficiency or cost for similar appliances that compete with types having similar dimensions and characteristics. A graph, or grade, comparing the designed model with other types is also necessary. Energy labels will work effectively when a variety of model efficiencies exist on the market. The purpose of introducing labels is to convince consumers to buy and manufacturers to produce more-efficient products. Three kinds of labels are distinguished:

- endorsement labels;
- comparative labels; and
- information-only labels.

Without extensive technical knowledge, energy-efficiency labels can offer a comparison of physical characteristics, related to energy-efficiency and environment compatibility, support consumers to make purchase decisions and enhance motivation for the manufacturer to develop and market energy-efficiency equipment.

When buying products according to the utilization status of energy-saving, it is hoped that there is more information to be provided for the product. However, traditionally only the basic data such as wattage value is available as an indicator of the maximum energy consumption. But this information is just one static value and it does not relate to operational and state reference parameters: there is neither a relation to a certain operating state (peak value or average value) of the product nor the degree of fulfilling functional and performance aspects. Against that, energy labels as 'informative labels affixed to manufactured products to describe the product's energy performance' are an established way to integrate complex energy-related information in an easy and comprehensible format. They inform consumers about specific energy consumption of the considered products and enable easy but accurate comparisons regarding energy-efficiency due to linking condition-based energy data with performance measures. For the last few years the numbers of applied energy labels have been increased worldwide. They differ regarding their underlying criteria, potential legal restrictions (e.g. mandatory labels versus voluntary labels), testing procedures, type of classification (endorsement label versus comparison label), design or evaluation monitoring procedures.

As mentioned above, one major focus of energy labels is on home appliances, computers and electronics, and lighting, but now the focus is moving towards industrial equipment and buildings (see Figure 3.3). Additionally, main drivers of energy consumption in industries are yet to be considered. Components such as electrical motors, which can account for up to two-thirds of electrical power consumption in industries, and pumps are notable examples here.

Labels and standards include a variety of policy instruments that play different roles in government efforts to encourage the development, marketing and sale of energy-efficient products. Labels are markings, with supporting promotion and directories, which show products' energy use or efficiency according to a common measure.

Comparison labels indicate the energy-efficiency of a particular model relative to similar models on the market, and are usually, though not always, mandatory. Energy-efficiency comparison labels are informative labels affixed to manufactured products to describe the product's energy

Figure 3.3 **Tendencies regarding the focus of present and future energy labels**

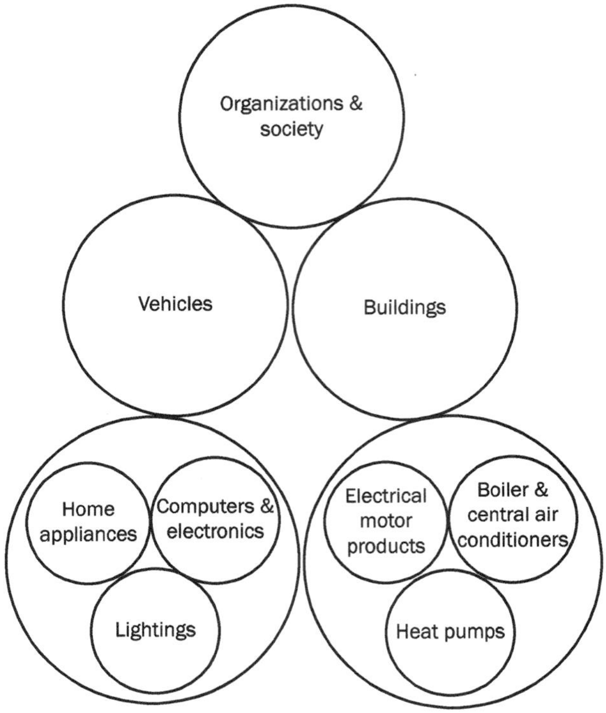

performance, usually in the form of energy use, efficiency or energy cost (see Figure 3.4). Labels shift the distribution of energy-efficient models upward by providing information that allows consumers to make rational decisions and stimulating manufacturers to design products that achieve higher ratings than the minimum standard. They usually allow consumers to compare performance among similar products using either discrete categories of performance or a continuous scale.

One of the most extensive and widely known programmes of **endorsement labels** is the US ENERGY STAR® programme. ENERGY STAR is a voluntary labelling

Figure 3.4 Comparison labels

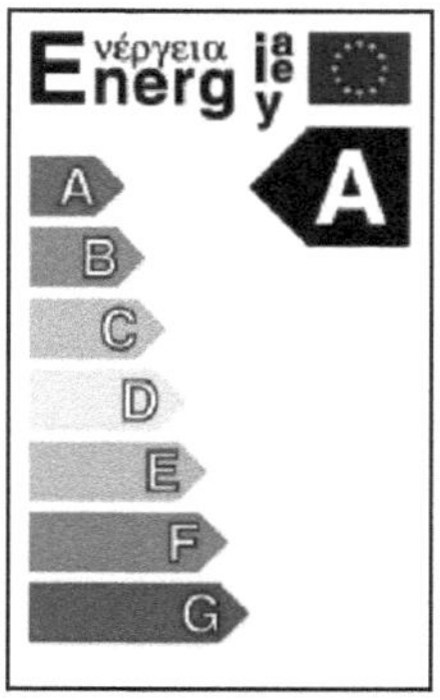

programme operated jointly by the United States Department of Energy (US DOE) and the United States Environmental Protection Agency (US EPA). It was introduced in 1992 to recognize energy-efficient computers. The ENERGY STAR endorsement labelling programme has grown to identify efficient products in more than 40 categories, including household appliances, home electronics (televisions, audio systems, etc.), computers and other office equipment, residential heating and cooling equipment, and lighting devices. ENERGY STAR standards also cover such products as distribution transformers, where the prospective purchasers are not households. The ENERGY STAR labels are shown in Figure 3.5.

Figure 3.5 Endorsement labels

Figure 3.6 Information-only label

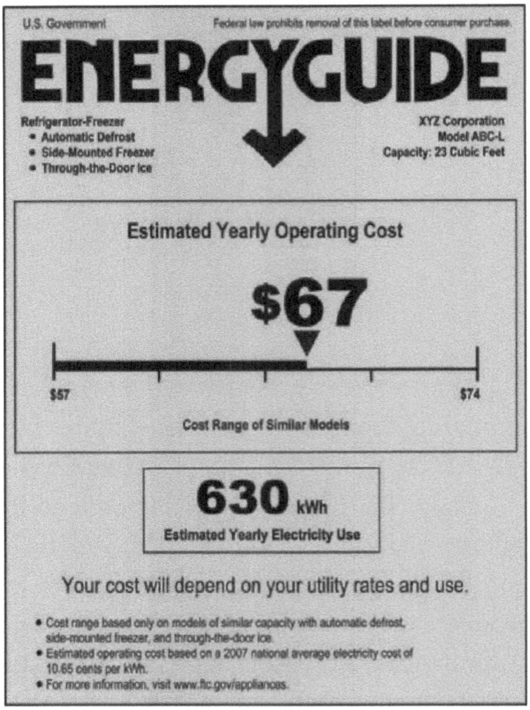

Information-only labels provide information on the technical performance of the single labelled product and offer no simple way (such as a ranking system) to compare energy performance between products. These types of label are generally not consumer friendly because they contain only technical information (see Figure 3.6).

3.5 Energy-efficiency testing methods

An energy test procedure is a standardized method for measuring the energy performance of an appliance or piece of equipment (Mahlia et al., 2003). A test procedure is a well-defined protocol or laboratory test method by which a relative ranking of the energy-efficiency among alternative technological designs that provide an energy-consuming service can be obtained. Energy test procedures lay the technical foundation for all energy-efficiency standards, labels and other related programmes. Their relationship is clearly clarified in Figure 3.1. The test procedure provides consistent measurements of appliance energy consumption.

3.6 Conclusion

Like the home appliances, the energy-efficiency of electrical motor products has been emphasized in recent years. Especially for electric motors, compulsory energy-efficiency regulations have come into effect in the EU, China and the USA. Energy-efficiency regulations, agreements, standards, labels and testing methods form the basis of energy-efficiency technical measurements for electrical motor products.

4

Energy-efficiency standards of electrical motor products

Abstract: The energy-efficiency standards for electric motors, pumps and fans are presented in this chapter. The mandatory energy-efficiency standards for electric motors in the USA, the EU and China are introduced. The mandatory and voluntary energy-efficiency requirements for pumps in China and the EU are discussed. The ISO and AMCA energy-efficiency standards for fans are presented, while the mandatory requirements for fans in China and the EU are also given.

Key words: energy-efficiency, standards, electric motor, pump, fan.

4.1 Electric motor

4.1.1 IEC 60034–30:2008

Many different energy-efficiency standards for induction electric motors are currently in use around the world. It becomes increasingly difficult for manufacturers to design motors for a global market and for customers to understand differences and similarities of standards in different countries. In order to overcome this undesirable situation, international standard IEC 60034–30:2008 was developed by the IEC, which is intended to globally harmonize motor energy-efficiency classes in general purpose.

The IEC 60034–30:2008 standard defines energy-efficiency classes for single-speed, three-phase, 50Hz and 60Hz cage-induction motors that:

- have a rated voltage (UN) up to 1,000V;
- have a rated output power (PN) between 0.75kW and 375kW;
- have either two, four or six poles;
- are rated on the basis of duty type S1 (continuous duty) or S3 (intermittent periodic duty) with an operation time of 80 per cent or more;
- are capable of operating direct online;
- are rated for operating conditions according to IEC 60034–1, clause 6.

Motor energy-efficiency has been classified based on their nominal load. Shown as Figure 4.1, IEC 60034–30:2008 defines three efficiency classes for 50 and 60Hz AC induction motors between 0.75kW and 375kW:

- IE3: premium efficiency (identical to NEMA premium);
- IE2: high efficiency (shown in Tables 4.1 and 4.2);
- IE1: standard efficiency (shown in Tables 4.1 and 4.2).

Figure 4.1 presents the IE efficiency classes for two-pole motors. Nominal limits (%) for high efficiency (IE2) and premium efficiency (IE3) at 50Hz and 60Hz are defined respectively in IEC 60034–30 ed.1.0 (see Tables 4.1 and 4.2, Copyright © 2008 IEC Geneva, Switzerland. www.iec.ch).

Efficiency and losses shall be tested in accordance with IEC 60034–2–1. For IE1, test methods associated with low and medium uncertainty are acceptable. The selected test method shall be stated in the documentation of the motor.

Figure 4.1 **Three efficiency classes defined by IEC 60034-30:2008**

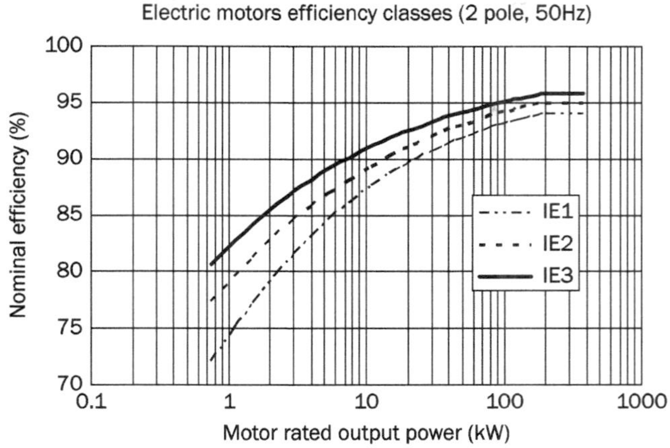

Table 4.1 **Nominal limits (%) for high efficiency (IE2) and premium efficiency (IE3) (50Hz) (Copyright © 2008 IEC Geneva, Switzerland. www.iec.ch)**

Rated output power (kW)	Standard efficiency (%)					
	IE2			IE3		
	2 poles	4 poles	6 poles	2 poles	4 poles	6 poles
0.75	77.4	79.6	75.9	80.7	82.5	78.9
1.1	79.6	81.4	78.1	82.7	84.1	81.0
1.5	81.3	82.8	79.8	84.2	85.3	82.5
2.2	83.2	84.3	81.8	85.9	86.7	84.3
3	84.6	85.5	83.3	87.1	87.7	85.6
4	85.8	86.6	84.6	88.1	88.6	86.8
5.5	87.0	87.7	86.0	89.2	89.6	88.0
7.5	88.1	88.7	87.2	90.1	90.4	89.1
11	89.4	89.8	88.7	91.2	91.4	90.3
15	90.3	90.6	89.7	91.9	92.1	91.2
18.5	90.9	91.2	90.4	92.4	92.6	91.7
22	91.3	91.6	90.9	92.7	93.0	92.2
30	92.0	92.3	91.7	93.3	93.6	92.9

(continued)

Table 4.1 Nominal limits (%) for high efficiency (IE2) and premium efficiency (IE3) (50Hz) (*cont'd*)

Rated output power (kW)	Standard efficiency (%)					
	IE2			IE3		
	2 poles	4 poles	6 poles	2 poles	4 poles	6 poles
37	92.5	92.7	92.2	93.7	93.9	93.3
45	92.9	93.1	92.7	94.0	94.6	94.1
55	93.2	93.5	93.1	94.3	94.6	94.1
75	93.8	94.0	93.7	94.7	95.0	94.6
90	94.1	94.2	94.0	95.0	95.2	94.9
110	94.3	94.5	94.3	95.2	95.4	95.1
132	94.6	94.7	94.6	95.4	95.6	95.4
160	94.8	94.9	94.8	95.6	95.8	95.6
200–375	95.0	95.1	95.0	95.8	96.0	95.8

Table 4.2 Nominal limits (%) for high efficiency (IE2) and premium efficiency (IE3) (60Hz) (Copyright © 2008 IEC Geneva, Switzerland. www.iec.ch)

Rated output power (kW)	Standard efficiency (%)					
	IE2			IE3		
	2 poles	4 poles	6 poles	2 poles	4 poles	6 poles
0.75	75.5	82.5	80.0	77.0	85.5	82.5
1.1	82.5	84.0	85.5	84.0	86.5	87.5
1.5	84.0	84.0	85.5	84.0	86.5	87.5
2.2	85.5	87.5	87.5	86.5	89.5	89.5
3.7	87.5	87.5	87.5	88.5	89.5	89.5
5.5	88.9	89.5	89.5	89.5	91.7	91.0
7.5	89.5	89.5	89.5	90.2	91.7	91.0
11	90.2	91.0	90.2	91.0	92.4	91.7
15	90.2	91.0	90.2	91.0	93.0	91.7
18.5	91.0	92.4	91.7	91.7	93.6	93.0
22	91.0	92.4	91.7	91.7	93.6	93.0
30	91.7	93.0	93.0	92.4	94.1	94.1

37	92.4	93.0	93.0	93.0	94.5	94.1
45	93.0	93.6	93.6	93.6	95.0	94.5
55	93.0	94.1	93.6	93.6	95.4	95.0
75	93.6	94.5	94.1	94.1	95.4	95.0
90	94.5	94.5	94.1	95.0	95.4	95.0
110	94.5	95.0	95.0	95.0	95.8	95.8
150	95.0	95.0	95.0	95.4	96.2	95.8
185–375	95.4	95.4	95.0	95.8	96.2	95.8

Source: The authors thank the International Electrotechnical Commission (IEC) for permission to reproduce information from its International Standard IEC 60034-30 ed.1.0 (2008). Further information on the IEC is available from www.iec.ch. IEC has no responsibility for the placement and context in which the extracts and contents are reproduced by the author, nor is IEC in any way responsible for the other content or accuracy therein.

For all higher energy-efficiency levels, only methods associated with low uncertainty shall be acceptable.

Normative, 50Hz, nominal limits for rated output power values not defined in the tables within the output power range of 0.75kW up to 200kW shall be computed by applying the equation (3.1). For general use the following formula may be applied:

$$\eta_N = A \cdot \left[\log_{10}\left(\frac{P_N}{1kW}\right)\right]^3 + B \cdot \left[\log_{10}\left(\frac{P_N}{1kW}\right)\right]^2 + C \cdot \left[\log_{10}\left(\frac{P_N}{1kW}\right)\right] + D \tag{3.1}$$

where *A, B, C, D* = interpolation coefficients (see Table 4.3); P_n is given in kW.

Normative, 60Hz, nominal limits for rated output power values not defined in Table 4.2 shall be determined as follows:

Table 4.3 **Interpolation coefficients (informative)**

IE-code	coefficients	50Hz, below 200kW		
		2 poles	4 poles	6 poles
IE1	A	0.5234	0.5234	0.0786
	B	–5.0499	–5.0499	–3.5838
	C	17.4180	17.4180	17.2918
	D	74.3171	74.3171	72.2383
IE2	A	0.2972	0.0278	0.0148
	B	–3.3454	–1.9247	–2.4978
	C	13.0651	10.4395	13.2470
	D	79.077	80.9761	77.5603
IE3	A	0.3569	0.0773	0.1252
	B	–3.3076	–1.8951	–2.613
	C	11.6108	9.2984	11.9963
	D	82.2503	83.7025	80.4769

- The efficiency of a rated power at or above the midpoint between two consecutive power ratings shall be the higher of the two efficiencies.
- The efficiency of a rated power below the midpoint between two consecutive power ratings shall be the lower of the two efficiencies.

4.1.2 NEMA MG1–2009

The National Electrical Manufacturers Association (NEMA), founded in 1926, is the leading trade association in the US representing the interests of electrical industry manufacturers. NEMA has been promoting standardization of electric motor efficiency since the 1970s and keep these standards up to date with the invention of new materials and designs.

NEMA MG1–2009 is produced by NEMA's Motor and Generator Section. A premium class is established by NEMA with rated and minimum efficiency levels for closed and open motors from two to six poles, and includes motors up to 500hp (National Electrical Manufacturers Association, 2009). Having successfully completed phase 2 – for motor efficiency improvement (phase 1 was the adoption of EPAct motors, see Table 4.4), NEMA is moving forward to phase 3, which focuses on motor systems efficiency improvements. This is a much more complex process and will provide

Table 4.4 General purpose motors required full-load nominal efficiency under EPACT-92

Motor horsepower	Nominal full-load efficiency					
	Open motors			Enclosed motors		
	6-pole	4-pole	2-pole	6-pole	4-pole	2-pole
1	80.0	82.5		80.0	82.5	75.5
1.5	84.0	84.0	82.5	85.5	84.0	82.5
2	85.5	84.0	84.0	86.5	84.0	84.0
3	86.5	86.5	84.0	87.5	87.5	85.5
5	87.5	87.5	85.5	87.5	87.5	87.5
7.5	88.5	88.5	87.5	89.5	89.5	88.5
10	90.2	89.5	88.5	89.5	89.5	89.5
15	90.2	91.0	89.5	90.2	91.0	90.2
20	91.0	91.0	90.2	90.2	91.0	90.2
25	91.7	91.7	91.0	91.7	92.4	91.0
30	92.4	92.4	91.0	91.7	92.4	91.0
40	93.0	93.0	91.7	93.0	93.0	91.7
50	93.0	93.0	92.4	93.0	93.0	92.4
60	93.6	93.6	93.0	93.6	93.6	93.0
75	93.6	94.1	93.0	93.6	93.6	93.0
100	94.1	94.1	93.0	94.1	94.5	93.6
125	94.1	94.5	93.6	94.1	94.5	94.5
150	94.5	95.0	93.6	95.0	95.0	94.5
200	94.5	95.0	94.5	95.0	95.0	95.0

significantly more energy savings. NEMA members felt this approach was currently more important than developing another level of 'super' premium efficiency for motors.

4.1.3 EISA 2007 - section 313

The Energy Independence and Security Act (EISA), which was passed by the US Congress and signed into law on 19 December 2007, became effective on 19 December 2010. The legislation updates the 1992 Energy Policy Act (EPAct), which covered low-voltage, general-purpose (subtype I), three-phase electric motors in the 1hp to 200hp range.

EISA expands to a wider range of industrial motors, from 1–500hp, which are manufactured for sale in the United States. The US Department of Energy (DOE) is responsible for establishing the rules to implement and enforce the EPAct. The motor categories in NEMA MG1 are as follows:

General purpose electric motor (subtype I)

This includes any electric motor that meets the definition of 'general purpose' motor as established by the DOE in its final rule codified in 10 CFR 431. For each general-purpose rating (subtype I) from 1–200hp that was previously covered by the EPAct, the law specifies a nominal full-load efficiency level based on NEMA premium efficiency as shown in NEMA MG 1-2009 Table 12–12 ('NEMA premium' efficiency levels) (see Table 4.5). All 230 or 460 volt (and 575 volts for Canada) motors manufactured after 19 December 2010 must meet or exceed this efficiency level. General-purpose electric motors (subtype II) not previously covered by the EPAct will be required to comply with energy efficiencies defined by NEMA MG 1 (2009).

Table 4.5 NEMA MG-1 Table 12-12 Full-load efficiencies for 60Hz NEMA Premium® efficient electric motors rated 600 volts or less (random wound) (Reprinted from [NEMA MG1-2009, Revision 1-2010] by permission of the National Electrical Manufacturers Association)

Motor horsepower	Nominal full-load efficiency					
	Open motors			Enclosed motors		
	2 pole	4 pole	6 pole	2 pole	4 pole	6 pole
1.0	77.0	85.5	82.5	77.0	85.5	82.5
1.5	84.0	86.5	86.5	84.0	86.5	87.5
2.0	85.5	86.5	87.5	85.5	86.5	88.5
3.0	85.5	89.5	88.5	86.5	89.5	89.5
5.0	86.5	89.5	89.5	88.5	89.5	89.5
7.5	88.5	91.0	90.2	89.5	91.7	91.0
10.0	89.5	91.7	91.7	90.2	91.7	91.0
15.0	90.2	93.0	91.7	91.0	92.4	91.7
20.0	91.0	93.0	92.4	91.0	93.0	91.7
25.0	91.7	93.6	93.0	91.7	93.6	93.0
30.0	91.7	94.1	93.6	91.7	93.6	93.0
40.0	92.4	94.1	94.1	92.4	94.1	94.1
50.0	93.0	94.5	94.1	93.0	94.5	94.1
60.0	93.6	95.0	94.5	93.6	95.0	94.5
75.0	93.6	95.0	94.5	93.6	95.4	94.5
100.0	93.6	95.4	95.0	94.1	95.4	95.0
125.0	94.1	95.4	95.0	95.0	95.4	95.0
150.0	94.1	95.8	95.4	95.0	95.8	95.8
200.0	95.0	95.8	95.4	95.4	96.2	95.8
250.0	95.0	95.8	95.4	95.4	96.2	95.8
300.0	95.4	95.8	95.4	95.4	96.2	95.8
350.0	95.4	95.8	95.4	95.4	96.2	95.8
400.0	95.8	95.8	95.8	95.4	96.2	95.8
450.0	95.8	96.2	96.2	95.4	96.2	95.8
500.0	95.8	96.2	96.2	95.4	96.2	95.8

General purpose electric motor (subtype II)

The general purpose electric motor (subtype II), manufactured after 10 December 2010, shall have a nominal full-load efficiency that is not less than as defined in NEMA MG 1–2009 Table 12–11 (see Table 4.6). Any electric motor incorporating the design elements of a general purpose electric motor (subtype I) into either of the following motor designs – U-frame motor, design-C, close-coupled pump, footless, vertical solid shaft normal thrust (tested in a horizontal configuration), eight-pole (900rpm), and polyphase 600V and below (other than 230 or 460V) – shall have a nominal full-load efficiency that is not less than as defined in NEMA MG 1–2009 Table 12–12.

Each NEMA Design B, general purpose electric motor with a power rating of 201–500hp, manufactured after 19 December 2010, which is not previously covered by the EPAct, will be required to comply with energy efficiencies as defined by NEMA MG 1–2006, Table 12–11.

Full-load nominal efficiency is determined using the same test methods as EPAct. EISA specifies that the test procedures for determining a motor's efficiency should follow IEEE standard 112, Test Method B or CSA 390.

4.1.4 EC EU regulation 640/2009

The EU officially published the framework of the ErP on 11 August 2005. Directive 2005/32/EC (Ecodesign Directive for Energy-rated Products – ErP) establishes a framework for the setting of general ecodesign requirements to be fulfilled by energy-using products with a view to increasing the energy-efficiency of the products and the level of environment protection. Within the framework of the Directive, specific regulations have been issued for the setting of design and energy-efficiency requirements, applicable to the products

Table 4.6 **NEMA MG1 Table 12–11 Full-load efficiencies of energy-efficiency motors (Reprinted from [NEMA MG1-2009, Revision 1-2010] by permission of the National Electrical Manufacturers Association)**

Motor horse-power	Nominal full-load efficiency							
	Open motors				Enclosed motors			
	2 pole	4 pole	6 pole	8 pole	2 pole	4 pole	6 pole	8 pole
1.0	N/A	82.5	80.0	74.0	75.5	82.5	80.0	74.0
1.5	82.5	84.0	84.0	75.5	82.5	84.0	85.5	77.0
2.0	84.0	84.0	85.5	85.5	84.0	84.0	86.5	82.5
3.0	84.0	86.5	86.5	86.5	85.5	87.5	84.5	84.0
5.0	85.5	87.5	87.5	87.5	87.5	87.5	87.5	85.5
7.5	87.5	88.5	88.5	88.5	88.5	89.5	89.5	85.5
10.0	88.5	89.5	90.2	89.5	89.5	89.5	89.5	88.5
15.0	89.5	91.0	90.2	89.5	90.2	91.0	90.2	88.5
20.0	90.2	91.0	91.0	90.2	90.2	91.0	90.2	89.5
25.0	91.0	91.7	91.7	90.2	91.0	92.4	91.7	89.5
30.0	91.0	92.4	92.4	91.0	91.0	92.4	91.7	91.0
40.0	91.7	93.0	93.0	91.0	91.7	93.0	93.0	91.0
50.0	92.4	93.0	93.0	91.7	92.4	93.0	93.0	91.7
60.0	93.0	93.6	93.6	92.4	93.0	93.6	93.6	91.7
75.0	93.0	94.1	93.6	93.6	93.0	94.1	93.6	93.0
100.0	93.0	94.1	94.1	93.6	93.6	94.5	94.1	93.0
125.0	93.6	94.5	94.1	94.5	94.5	94.5	94.1	93.6
150.0	93.6	95.0	94.5	94.5	94.5	95.0	95.0	93.6
200.0	94.5	95.0	94.5	95.0	95.0	95.0	95.0	94.1
250.0	94.5	95.4	95.4	95.4	95.4	95.0	95.0	94.5
300.0	95.0	95.4	95.4	—	95.4	95.0	95.0	—
350.0	95.0	95.4	95.4	—	95.4	95.4	95.0	—
400.0	95.4	95.4	—	—	95.4	95.4	—	—
450.0	95.8	95.8	—	—	95.4	95.4	—	—
500.0	95.8	95.8	—	—	95.4	95.8	—	—

included in the scope of the Directive. Among these, Regulation 640/2009 EC establishes ecodesign requirements for electric motors and their drives in terms of energy-efficiency levels.

Regulation 640/2009 EC is applicable to electric motors, the type and size of which are defined in Article 2.1: single-speed, three-phase, squirrel-cage induction motors, with two, four or six poles, rated voltage up to 1,000V, rated power output between 0.75 to 375kW (rating based on continuous duty). Efficiency levels in the Regulation are in accordance with IEC 60034–30:2008 and in compliance with IEC 60034–2–1:2007. The motor energy-efficiency is classified into two levels: IE2 and IE3. Ecodesign requirements for electric motors shall be applied in accordance to the following timetable:

- from 16 June 2011: motors shall not be less efficient than the IE2 level;
- from 1 January 2015: motors with a rated output of 7.5–375kW shall not be less efficient than IE3 or meet IE2, if equipped with a variable speed drive;
- from 1 January 2017: all motors with a rated power of 0.75–375kW shall not be less efficient than IE3 or meet IE2, if equipped with a variable speed drive.

The energy-efficiency levels adopted by the EU and the US for electric motors are compared in Figure 4.2.

4.1.5 GB 18613-2006

GB 18613 is the mandatory energy-efficiency standard for electric motors in China to promote the energy-efficiency improvement of the motor and motor system. China has mandatory ratings for standard motors and voluntary ratings for high-efficiency motors from 0.75–420hp (0.55–315kW), and for two to six poles (Standardization Administration of

Figure 4.2 **The energy-efficiency levels adopted by the EU and US**

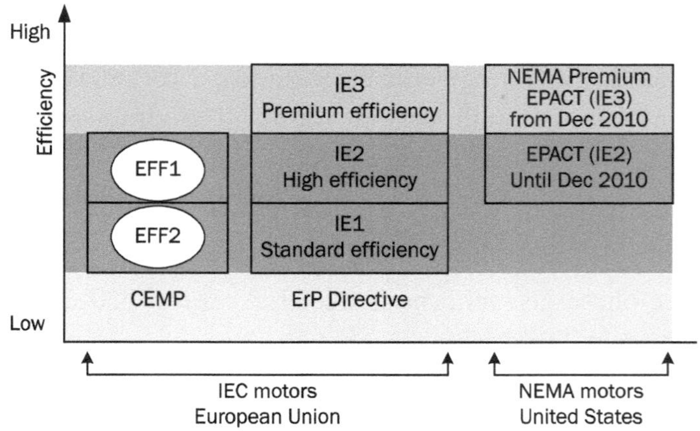

China, 2006). The EFF1 and EFF2 efficiency classification of the EU-CEMEP motor efficiency voluntary agreement was directly adopted as grade 2 and grade 3 respectively. The most efficient value, grade 1, was developed based on the Australian motor standard AS/NZ 1359.5. Standard GB18613–2006 will be revised according to the motor efficiency test standard IEC1 60034–2–1 and the first global energy-efficiency classification standard IEC 60034–30. The efficiency classification in the new version of GB 18613 will be adjusted: grade 2 from the current rating system will become grade 3 in the new rating system, thus the current grade 3 will be eliminated. The new grade 3 will directly be equal to the IE2 classification and grade 3 to IE3. The highest efficiency grade 1 will be based on IE3 with 15 per cent loss reduction (IE4) and also refer to related information specified in IEC 60034–31. The new standard shall be issued by the Chinese government in 2011.

Although the stray losses P_{LL} are regarded as 0.5 per cent input power in GB 18613–2006, the new method with summation of losses with load test will be adopted: the P_{LL} will be determined from residual loss.

4.1.6 Minimum allowable values of energy-efficiency and values of efficiency grade for small-power motors

The mandatory energy-efficiency standard for small-power motors will be published in China. The draft standard for approval defines energy-efficiency classes for a rated voltage up to 690V, 50Hz, AC supply (Standardization Administration of China, 2010):

- three-phase asynchronous electric motor (10–2200W) (shown in Table 4.7);

Table 4.7 Nominal minimum efficiency for small-power three-phase electric motors

Rated power (W)	Energy-efficiency classes								
	Level 1			Level 2			Level 3		
	2 poles	4 poles	6 poles	2 poles	4 poles	6 poles	2 poles	4 poles	6 poles
10	–	35.0	–	–	31.4	–	–	28.0	–
16	54.1	39.4	–	50.1	35.6	–	46.0	32.0	–
25	60.0	50.1	–	56.0	46.0	–	52.0	42.0	–
40	62.8	58.1	–	59	54.1	–	55	50	–
60	67.5	63.8	–	63.8	60	–	60	56	–
90	69.3	65.7	–	65.7	61.9	–	62	58	–
120	73.8	67.5	–	70.5	63.8	–	67	60	–
180	75.5	71.1	66.6	72.4	67.7	62.9	69	64	59
250	78.1	73.8	70.2	75.2	70.5	66.7	72	67	63
370	79.3	75.9	74.6	76.5	72.8	71.4	73.5	69.5	68
550	81.0	79.3	77.2	78.4	76.5	74.2	75.5	73.5	71
750	–	–	–	–	–	–	77.4	79.6	75.9
1100	–	–	–	–	–	–	79.6	81.4	78.1
1500	–	–	–	–	–	–	81.3	82.8	79.8
2200	–	–	–	–	–	–	83.2	84.3	81.8

- capacitor-run asynchronous electric motor (10–2,200W) (shown in Table 4.8);
- capacitor-start asynchronous electric motor (120–3,700W) (shown in Table 4.9);
- two-value capacitor asynchronous electric motor (250–3,000W) (shown in Table 4.10);
- electric motors used in air conditioner to drive fans (6–550W) (shown in Table 4.11).

According to the above analysis, the energy-efficiency directive for electric motors adopted in the US, the EU and China is summarized Table 4.12.

Table 4.8 Nominal minimum efficiency for capacitor-run electric motors

Rated power (W)	Energy-efficiency classes								
	Level 1			Level 2			Level 3		
	2 poles	4 poles	6 poles	2 poles	4 poles	6 poles	2 poles	4 poles	6 poles
10	–	30.4	–	–	27.1	–	–	24	–
16	42.7	40.5	–	38.8	36.7	–	35	33	–
25	48	45.9	–	44	41.9	–	40	38	–
40	55.1	53.1	–	51.1	49	–	47	45	–
60	60.9	58.1	–	57	54.1	–	53	50	–
90	63.8	60	–	60	56	–	56	52	–
120	67.5	64.7	60.9	63.8	60.9	57	60	57	53
180	72	66.6	62.8	68.6	62.9	59	65	59	55
250	72.9	68.4	64.7	69.5	64.8	60.9	66	61	57
370	73.8	69.3	65.7	70.5	65.7	61.9	67	62	58
550	76.4	71.1	69.3	73.3	67.7	65.7	70	64	62
750	78.1	74.6	72	75.2	71.4	68.6	72	68	65
1100	80.6	77.2	–	77.9	74.2	–	75	71	–
1500	81.4	78.9	–	78.8	76.1	–	76	73	–
2200	82.2	–	–	79.8	–	–	77	–	–

Table 4.9 Nominal minimum efficiency for capacitor-start electric motors

Rated power (W)	Energy-efficiency classes								
	Level 1			Level 2			Level 3		
	2 poles	4 poles	6 poles	2 poles	4 poles	6 poles	2 poles	4 poles	6 poles
120	–	58.1	–	–	54.1	–	–	50.0	–
180	67.5	60.9	–	63.8	57.0	–	60.0	53.0	–
250	71.1	65.7	61.9	67.7	61.9	58.0	64.0	58.0	54.0
370	72.0	69.3	65.7	68.6	65.7	61.9	65.0	62.0	58.0
550	74.6	72.9	67.5	71.4	69.5	63.8	68.0	66.0	60.0
750	76.4	74.6	68.4	73.3	71.4	64.8	70.0	68.0	61.0
1100	78.1	77.2	70.2	75.2	74.2	66.7	72.0	71.0	63.0
1500	79.8	78.9	74.6	77.0	76.1	71.4	74.0	73.0	68
2200	80.6	79.8	76.4	77.9	77.0	73.3	75.0	74.0	70
3000	81.4	80.6	–	78.8	77.9	–	76.0	75.0	–
3700	82.2	81.4	–	79.8	78.8	–	77.0	76.0	–

Table 4.10 Nominal minimum efficiency for two-value capacitor-start electric motors

Rated power (W)	Energy-efficiency classes					
	Level 1		Level 2		Level 3	
	2 poles	4 poles	2 poles	4 poles	2 poles	4 poles
250	–	69.3	–	65.7	–	62.0
370	73.8	72.0	70.5	68.6	67.0	65.0
550	76.4	74.6	73.3	71.4	70.0	68.0
750	78.1	77.2	75.2	74.2	72.0	71.0
1100	80.6	78.9	77.9	76.1	75.0	73.0
1500	81.4	80.6	78.8	77.9	76.0	75.0
2200	82.2	81.4	79.8	78.8	77.0	76.0
3000	83.9	82.2	81.6	79.8	79.0	77.0

Table 4.11 Nominal minimum efficiency for electric motor used in air conditioner to drive fans

Rated power (W)	Energy-efficiency classes								
	Level 1			Level 2			Level 3		
	4 poles	6 poles	8 poles	4 poles	6 poles	8 poles	4 poles	6 poles	8 poles
6	20.9	18.4	17.1	18.3	16.1	15.0	16.0	14.0	13.0
10	29.2	25.7	24.5	26.0	22.7	21.6	23.0	20.0	19.0
16	36.1	32.7	30.4	32.5	29.2	27.1	29.0	26.0	24.0
20	39.4	36.1	32.7	35.6	32.5	29.2	32.0	29.0	26.0
25	43.8	39.4	36.1	39.8	35.6	32.5	36.0	32.0	29.0
35	48.0	43.8	40.5	44.0	39.8	36.7	40.0	36.0	33.0
40	50.1	45.9	42.7	46.0	41.9	38.8	42.0	38.0	35.0
50	53.1	49.0	44.8	49.0	45.0	40.9	45.0	41.0	37.0
60	55.1	51.1	46.9	51.1	47.0	42.9	47.0	43.0	39.0
75	56.1	53.1	48.0	52.1	49.0	44.0	48.0	45.0	40.0
90	57.1	54.1	49.0	53.1	50.1	45.0	49.0	46.0	41.0
100	58.1	55.1	50.1	54.1	51.1	46.0	50.0	47.0	42.0
120	60.0	56.1	51.1	56.0	52.1	47.0	52.0	48.0	43.0
150	61.9	57.1	53.1	58.0	53.1	49.0	54.0	49.0	45.0
180	63.8	58.1	54.1	60.0	54.1	50.1	56.0	50.0	46.0
250	66.6	60.9	56.1	62.9	57.0	52.1	59.0	53.0	48.0
370	72.0	62.8	57.1	68.6	59.0	53.1	65.0	55.0	49.0
400	72.9	65.7	60.0	69.5	61.9	56.0	66.0	58.0	52.0
450	73.8	67.5	61.9	70.5	63.8	58.0	67.0	60.0	54.0
480	74.6	68.4	62.8	71.4	64.8	59.0	68.0	61.0	55.0
500	75.5	69.3	63.8	72.4	65.7	60.0	69.0	62.0	56.0
550	76.4	70.2	64.7	73.3	66.7	60.9	70.0	63.0	57.0

4.2 Pumps

4.2.1 GB 19762–2007

A mandatory Chinese Standard GB 19762–2007, named 'The minimum allowable values of energy efficiency and

Table 4.12 **The energy-efficiency directive for electric motors adopted in the US, EU and China**

Country	Directive/ Standard	Effective date	Efficiency level	Power	Voltage/ Frequency	Poles
USA	Epact 2007 EISA (NEMA MG1) (NEMA MG 1-2009)	Dec-10	Subtype I – NEMA premium (IE3)	1 up to 200hp	No more than 600v/ 60Hz	2 to 8
			Subtype II – high efficiency (IE2)	1 up to 200hp		
			NEMA Design B – high efficiency (IE2)	200 up to 500hp		
EU	640/2009 (IEC 60034-30)	Jun-11	High efficiency (IE2)	0.75 up to 375kW	Up to 1000V, 50 or 50/60 Hz	2 to 6
		Jan-15	Premium efficiency (IE3)	7.5 up to 375kW		
		Jan-17	Premium efficiency (IE3)	0.75 up to 375kW		
China	GB 18613-2006	Jul-11	High efficiency (IE2)	0.55 up to 315 kW	Up to 690V, 50Hz	2 to 6

evaluating values of energy conservation of centrifugal pump for fresh water', came into effect in July 2008. The standard gives a graph of efficiency against specific flow for single-stage (single and double suction) and multistage centrifugal pumps for clean water. When the flow rate is between 5m^3/h and 10,000m^3/h, the energy-efficiency limited value is determined in the energy-efficiency curve according to the specific speed (N/s) and pump flow rate Q (m^3/h). When the flow rate is bigger than 10,000m^3/h, as for a single-stage end suction centrifugal pump, the minimum allowable value of energy-efficiency is 87 per cent; for a double suction design centrifugal pump, the minimum allowable value of energy efficiency is 86 per cent.

4.2.2 European guide to pump efficiency for single-stage centrifugal pumps

The purpose of the European Guide to pump efficiency for single-stage centrifugal pumps was to help purchasers choose pumps of good efficiency. This guide should only be used for single-stage centrifugal pumps handling clean water at up to 95°C. Pump types covered are:

- end suction pumps having their own bearings;
- end suction close-coupled pumps, with the impeller on the motor shaft;
- double-entry pumps having an axially split casing.

It shows six plots of pump efficiency against flow, for end suction close coupled, end suction own bearings and double-entry split casing pumps, running at two- and four-pole speeds. Two lines were derived for each plot. The upper line represents the mean of the catalogue best efficiencies and is ideally the efficiency a user should aspire to for their pump main duty. However, since it is not always possible to source the ideal pump, another line was added, five to ten points below the upper line, to cover efficient pumps for which the required duty is away from the best efficiency point. Selection below the lower line was considered unacceptable unless there were exceptional circumstances. The applicable ranges

Figure 4.3 **Efficiencies of end suction pumps with their own bearings at 2,900rpm**

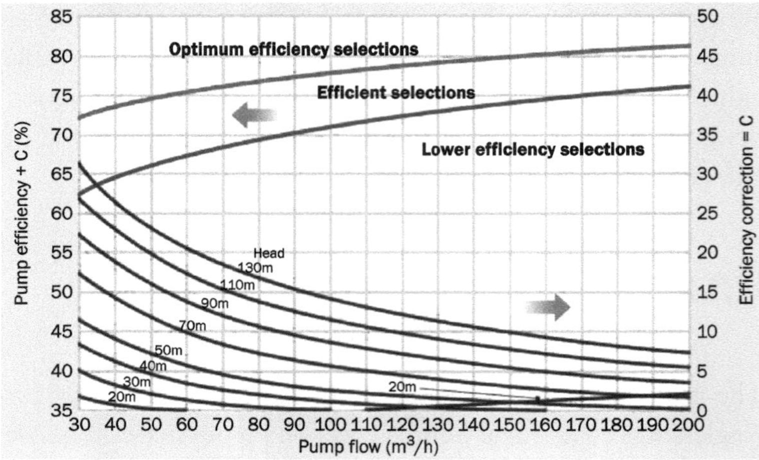

of flow and head are shown as Figure 4.3 for pumps running at nominally 2,900 and 1,450rpm (Europump, 2003).

4.2.3 (EC) No 641/2009

The ecodesign requirements for circulators are set out in (EC) No 641/2009 by the EU from 1 January 2013. The Regulations include:

- The energy-efficiency index of circulators, calculated in accordance with (EC) No 641/2009, shall be indicated on the name plate and packaging of the product and in the technical documentation as follows: EEI $\leqslant$ 0.27.
- The following information shall be provided: the benchmark for most efficient circulators is EEI $\leqslant$ 0.20.
- Information concerning disassembly, recycling or disposal at the end of life of components and materials shall be available for treatment facilities.

- The following information shall be provided on the packaging and in the technical documentation of drinking water circulators: 'this circulator is suitable for drinking water only.'

4.3 Fans

4.3.1 AMCA standard 205–10

The Air Movement and Control Association International (AMCA) has developed a standard (AMCA 205–10, Energy Efficiency Classification for Fans) that defines fan efficiency grades (FEG) based on the peak value of total efficiency of the fan without consideration of the motor/drive (The Air Movement and Control Association, 2007). This standard also recommends that any specification or code that sets a minimum acceptable FEG also include a requirement that the efficiency at the actual operating point(s) be within ten points of the peak value.

As AMCA spread outside of North America and became AMCA International, so too did the acceptance of AMCA-certified performance of fans in other regions of the world, including Europe, Asia and India. Furthermore, the AMCA Fan Committee established a subcommittee for developing ANSI/AMCA Standard 205 on fan energy-efficiency. This standard will be harmonized with the new ISO Standard 12759. AMCA STANDARD 205–10, Energy Efficiency Classification for Fans, was approved by AMCA International membership on 19 February 2010. It defines classification for all fan types designed to be driven by motors of nominal rating 125W (1/6hp) and above. Fans can range from purpose-built single fans to series-produced fans manufactured in large quantities. The standard also excludes classification for circulating fans.

4.3.2 ISO 12759:2010

The ISO technical committee ISO/TC 117, Fans, produced their own standard-ISO 12759: Efficiency classification for fans. As concerns the future ISO 12759, national differences or priorities have been taken into account. For instance, in Europe, the requirement is mostly for motor and fan impeller combinations, whereas in the USA the majority of fans are sold as bare shaft fans with the motor or drive mechanism being supplied separately. The ISO standard sets fan efficiency grades (FEG) for both bare shaft fans and fan-motor for driven fans.

Anyway, improving the efficiency of fans is the focus of a concerted effort now, both in the United States and worldwide. Standards are being written defining fan efficiency grades (FEG) and fan/motor efficiency grades (FMEG). ISO 12759:2010 specifies requirements for classification of fan efficiency for all fan types driven by motors with an electrical input power range from 0.125kW to 500kW. It is applicable to (bare shaft and driven) fans, as well as fans integrated into products. Fans integrated into products can be measured as stand-alone fans.

The peak efficiency that can be achieved in any fan is a function of fan type and size. Table 4.13 shows values of peak total efficiency achievable by production units with diameters of 24in. (600mm) or greater. The values in this table were generated from a thorough review of published catalogues from many US and EU manufacturers.

4.3.3 EuP Lot 11

The EU Commission has carried out a preparatory study which analysed the technical, environmental and economic aspects of fans in EuP Lot 11: Fans for ventilation in non-residential buildings. A regulation with regard to ecodesign

Table 4.13 **Peak total efficiency by fan type**

	Fan type	Peak efficiency range (%)
Centrifugal	Airfoil	88
	Backward curved	64
	Backward inclined	79–83
	Forward curved	65–70
Axial	Vane axial	78–86
	Tube axial	67–75
	Propeller	40–55
	Mixed flow	75
	Tangential	25

requirements for fans driven by motors with an electric input power between 125W and 500kW will be approved by the EU Commission. First tier requirements will apply on 1 January 2013 and the second tier requirements from 1 January 2015. The minimum efficiency requirements for fans are set out in Table 4.14 (http://ec.europa.eu/@European Union, 1995–2011).

The regulation includes fans used in many air-handling products (in particular ventilation). Fierce negotiations seem to be going on finalizing the exact scope of the measure, and in particular to decide whether built-in fans in larger products will be covered or not by the ecodesign rule. The measure should set minimum energy-efficiency requirements for several categories of fans and is expected to save 34TWh of electricity per year by 2020.

In Lot 11 report, eight different fan categories have been selected:

- axial fans of static pressure < 300 Pa;
- axial fans of static pressure > 300 Pa;
- centrifugal fans with forward curved (FC) blades;

Table 4.14 Minimum energy-efficiency requirements for fans (Used with permission of Publications Office of the European Union)

Fan types	Measurement category (A–D)	Efficiency category (static or total)	Target energy-efficiency	Efficiency grade	
				First tier	Second tier
Axial fans	A, C	Static	$\eta_{t\,arg\,et} = 2.74 \times \ln(p) - 6.33 + N$ $\eta_{t\,arg\,et} = 0.78 \times \ln(p) - 1.88 + N$	36	40
	B, D	Total	$\eta_{t\,arg\,et} = 2.74 \times \ln(p) - 6.33 + N$ $\eta_{t\,arg\,et} = 0.78 \times \ln(p) - 1.88 + N$	50	58
Centrifugal forward curved fan and centrifugal radial bladed fan	A, C	Static	$\eta_{t\,arg\,et} = 2.74 \times \ln(p) - 6.33 + N$ $\eta_{t\,arg\,et} = 0.78 \times \ln(p) - 1.88 + N$	37	44
	B, D	Total	$\eta_{t\,arg\,et} = 2.74 \times \ln(p) - 6.33 + N$ $\eta_{t\,arg\,et} = 0.78 \times \ln(p) - 1.88 + N$	42	49
Centrifugal backward curved fan without housing	A, C	Static	$\eta_{t\,arg\,et} = 4.56 \times \ln(p) - 10.5 + N$ $\eta_{t\,arg\,et} = 1.1 \times \ln(p) - 2.6 + N$	58	62
Centrifugal backward curved fan with housing	A, C	Static	$\eta_{t\,arg\,et} = 4.56 \times \ln(p) - 10.5 + N$ $\eta_{t\,arg\,et} = 1.1 \times \ln(p) - 2.6 + N$	58	61
	B, D	Total	$\eta_{t\,arg\,et} = 4.56 \times \ln(p) - 10.5 + N$ $\eta_{t\,arg\,et} = 1.1 \times \ln(p) - 2.6 + N$	61	64

Mixed flow fan	A, C	Static	$\eta_{t\,arg\,et} = 4.56 \times \ln(p) - 10.5 + N$ $\eta_{t\,arg\,et} = 1.1 \times \ln(p) - 2.6 + N$	47	50
	B, D	Total	$\eta_{t\,arg\,et} = 4.56 \times \ln(p) - 10.5 + N$ $\eta_{t\,arg\,et} = 1.1 \times \ln(p) - 2.6 + N$	58	62
Cross flow fan	B, D	Total	$\eta_{t\,arg\,et} = 1.14 \times \ln(p) - 2.6 + N$ $\eta_{t\,arg\,et} = N$	13	21
Box fan with backward curved or mixed flow fan	B, D	Total	$\eta_{t\,arg\,et} = 4.56 \times \ln(p) - 10.5 + N$ $\eta_{t\,arg\,et} = 1.1 \times \ln(p) - 2.6 + N$	35	39
Box fan with forward curved or axial fan	B, D	Total	$\eta_{t\,arg\,et} = 4.56 \times \ln(p) - 10.5 + N$ $\eta_{t\,arg\,et} = 1.1 \times \ln(p) - 2.6 + N$	22	25
Roof fan (axial fan within)	A, C	Static	$\eta_{t\,arg\,et} = 2.74 \times \ln(p) - 6.33 + N$ $\eta_{t\,arg\,et} = 0.78 \times \ln(p) - 1.88 + N$	27	31
Roof fan (centrifugal or mixed flow fan within)	A, C	Static	$\eta_{t\,arg\,et} = 4.56 \times \ln(p) - 10.5 + N$ $\eta_{t\,arg\,et} = 1.1 \times \ln(p) - 2.6 + N$	48	52

- centrifugal fans with backward curved (BC) blades without casing;
- centrifugal fans with backward curved (BC) blades with scroll housing;
- box fans;
- roof fans;
- cross-flow fans.

While in Commission Regulation (EU) No 327/2011, the box fans and roof fans are not considered

4.3.4. GB19761–2009

The Chinese standard GB19761–2009 is named 'Minimum allowable values of energy efficiency and energy efficiency grades for fan'. In this standard the energy-efficiency grade, minimum allowable values of energy-efficiency, evaluating values of energy-efficiency and testing methods are specified. The standard applies to general purpose centrifugal fans and axial flow fans. However, the standard is not applicable to jet fans and cross flow fans.

4.4 Conclusion

Energy-efficiency standards can be either mandatory or voluntary in nature. Nobody believes that energy-efficiency targets will remain constant. Both users who have to pay for the energy to run the fans and legislators who want to push the boundaries will be looking for incremental increases in efficiency levels.

For electric motors, the mandatory energy-efficiency standards must be compatible in the USA, the EU and China.

Table 4.15 The energy-efficiency grade for centrifugal fans

Press coefficient ϕ	Speed ratio n_s		efficiency η_r/%								
			No2<Machine No. <No5			No5 ≤Machine No. <No 10			Machine No. ≥No10		
			grade3	grade2	grade1	grade3	grade2	grade1	grade3	grade2	grade1
1.4~1.5	$45<n_s\leq65$		55	61	64	59	65	68	–	–	–
1.1~1.3	$35<n_s\leq45$		59	65	68	63	69	72	–	–	–
1.0	$10\leq n_s<20$		63	69	72	66	72	75	69	75	78
	$20\leq n_s<30$		65	71	74	68	74	77	71	77	80
0.9	$5\leq n_s<15$		66	72	75	69	75	78	72	78	81
	$15\leq n_s<30$		68	74	77	71	77	80	74	80	83
	$30\leq n_s<45$		70	76	79	73	79	82	76	82	85
0.8	$5\leq n_s<15$		66	72	75	69	75	78	72	78	81
	$15\leq n_s<30$		69	75	78	72	78	81	75	81	84
	$30\leq n_s<45$		71	77	80	74	80	83	76	82	85
0.7	$10\leq n_s<30$		68	74	77	70	76	79	72	79	83
	$30\leq n_s<50$		70	76	79	72	78	81	74	81	84
0.6	$20\leq n_s<45$	Airfoil	72	77	80	74	79	82	76	82	85
		Plate blade	69	74	77	71	76	79	73	79	83
	$45\leq n_s<70$	Airfoil	73	78	81	75	80	83	77	83	86
		Plate blade	70	75	78	72	77	80	74	80	83

(continued)

Table 4.15 The energy-efficiency grade for centrifugal fans (*cont'd*)

<table>
<tr><td rowspan="3">Press coefficient ϕ</td><td rowspan="3" colspan="2">Speed ratio n_s</td><td colspan="12">efficiency η_r/%</td></tr>
<tr><td colspan="3">No2<Machine No. <No5</td><td colspan="3">No5 ≤Machine No. <No 10</td><td colspan="6">Machine No. ≥No10</td></tr>
<tr><td>grade3</td><td>grade2</td><td>grade1</td><td>grade3</td><td>grade2</td><td>grade1</td><td colspan="2">grade3</td><td colspan="2">grade2</td><td colspan="2">grade1</td></tr>
<tr><td rowspan="6">0.5</td><td rowspan="2">$10 \leq n_s < 30$</td><td>Airfoil</td><td>70</td><td>76</td><td>79</td><td>72</td><td>78</td><td>81</td><td colspan="2">74</td><td colspan="2">81</td><td colspan="2">84</td></tr>
<tr><td>Plate blade</td><td>67</td><td>73</td><td>76</td><td>69</td><td>75</td><td>78</td><td colspan="2">71</td><td colspan="2">78</td><td colspan="2">81</td></tr>
<tr><td rowspan="2">$30 \leq n_s < 50$</td><td>Airfoil</td><td>73</td><td>79</td><td>82</td><td>75</td><td>81</td><td>84</td><td colspan="2">77</td><td colspan="2">83</td><td colspan="2">86</td></tr>
<tr><td>Plate blade</td><td>70</td><td>76</td><td>79</td><td>72</td><td>77</td><td>80</td><td colspan="2">74</td><td colspan="2">81</td><td colspan="2">84</td></tr>
<tr><td rowspan="2">$50 \leq n_s < 70$</td><td>Airfoil</td><td>75</td><td>80</td><td>83</td><td>77</td><td>82</td><td>85</td><td colspan="2">79</td><td colspan="2">84</td><td colspan="2">87</td></tr>
<tr><td>Plate blade</td><td>72</td><td>77</td><td>80</td><td>74</td><td>79</td><td>82</td><td colspan="2">76</td><td colspan="2">81</td><td colspan="2">84</td></tr>
<tr><td rowspan="6">0.4</td><td rowspan="2">$50 \leq n_s < 65$</td><td>Airfoil</td><td>76</td><td>81</td><td>84</td><td>78</td><td>83</td><td>86</td><td colspan="2">80</td><td colspan="2">85</td><td colspan="2">88</td></tr>
<tr><td>Plate blade</td><td>73</td><td>78</td><td>81</td><td>75</td><td>80</td><td>83</td><td colspan="2">77</td><td colspan="2">82</td><td colspan="2">85</td></tr>
<tr><td rowspan="4">$65 \leq n_s < 80$</td><td rowspan="2"></td><td colspan="3">Machine No. <No3.5</td><td colspan="3">No3.5 <Machine No.< No5</td><td>–</td><td>–</td><td>–</td><td>–</td><td>–</td><td>–</td></tr>
<tr><td>grade3</td><td>grade2</td><td>grade1</td><td>grade3</td><td>grade2</td><td>grade1</td><td>–</td><td>–</td><td>–</td><td>–</td><td>–</td><td>–</td></tr>
<tr><td>Airfoil</td><td>70</td><td>75</td><td>78</td><td>75</td><td>80</td><td>83</td><td>78</td><td>84</td><td>87</td><td>81</td><td>86</td><td>89</td></tr>
<tr><td>Plate blade</td><td>67</td><td>72</td><td>75</td><td>72</td><td>77</td><td>80</td><td>75</td><td>81</td><td>84</td><td>78</td><td>83</td><td>86</td></tr>
<tr><td rowspan="2">0.3</td><td rowspan="2">$65 \leq n_s < 80$</td><td>Airfoil</td><td colspan="3">–</td><td colspan="3">–</td><td>76</td><td>81</td><td>84</td><td>78</td><td>83</td><td>86</td></tr>
<tr><td>Plate blade</td><td colspan="3">–</td><td colspan="3">–</td><td>73</td><td>78</td><td>81</td><td>75</td><td>80</td><td>83</td></tr>
</table>

For pumps, the mandatory energy-efficiency standards have come into effect, while the ErP measure for pumps is studied and does not exist in the related standard in the USA. For fans, the AMCA has established an energy-efficiency standard – that is, AMCA 205–10 – and the ErP measure – Commission Regulation (EU) No 327/2011 was passed in EU Committee on March 2011. In China, the mandatory energy-efficiency standard has come into effect for pumps.

5

Measurement errors and uncertainties

Abstract: This chapter introduces the definition of measurement errors and uncertainties. The accuracy of measurement is described by trueness and precision. The uncertainties in the measurement are classified by type A and type B. The sources of uncertainties in measurements are explained. Finally, the evaluation process of uncertainties is described in detail.

Key words: measurement, errors, uncertainties, evaluation.

5.1 Classification of measurement errors

Measurement error is the difference between the true value and the measured value. In general, the result of a measurement is only an approximation or estimate of the value of the measurand and thus is complete only when accompanied by a statement of the uncertainty of the estimate.

The necessary components of any measurement are the method of measurement and the measuring instrument; measurements are often performed with the participation of a person (Rabinovich, 1999). The imperfection of each component of measurement contributes to the measurement error. There are three different kinds of errors (Fink, 2004):

1. Systematic errors either overestimate or underestimate the results of measurements and they arise for specific reasons, such as the incorrect set-up of the measuring equipment. As a consequence, the measurement process is affected and altered in one direction. From a strict theoretical point of view, systematic errors, different from random ones, can be fully compensated and their effect can be eliminated. The systematic errors have effects on the trueness of the measuring results.
2. Random errors arise from unforeseen events on the measurement. The theory of error is focused only on the study of gross and random errors. The random error cannot be compensated by applying a correction, but it can be effectively reduced by increasing the number of measurements. Random errors have effects on the precision of the measuring results.
3. Gross errors come from miscalculation of the data or from the incorrect reading of the measurement equipment. This kind of error is easy to identify because the measurement result differs from other results done before.

For this reason, in the general form, the total error in a scientific measurement can be expressed as:

$$\xi = \xi_s + \xi_r + \xi_g \tag{5.1}$$

Where ξ is the measurement error, and ξ_s, ξ_r, ξ_g are the systematic, random and gross errors, respectively.

The distinction between systematic and random errors is shown graphically in Figure 5.1. The scatter in the data represents random error and the deviation between the average of the readings and true value demonstrates the systematic error (Wheeler and Ganji, 2009).

Figure 5.1 Distinction between systematic and random errors

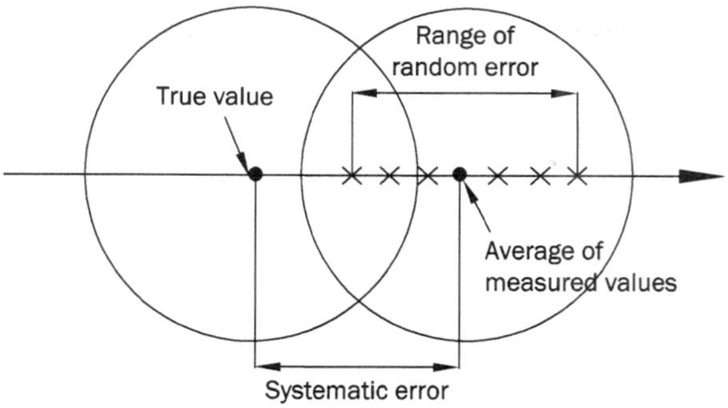

5.2 Trueness and precision

ISO 5725 deals with accuracy of measurement methods and results. It uses two terms – trueness and precision – to describe the accuracy of a measurement method (Deldossi and Zappa, 2009). **Trueness** refers to the closeness of agreement between the arithmetic mean of a large number of test results and the accepted reference value, while **precision** refers to the closeness of agreement between test results (Menditto et al., 2007).

Table 5.1 Definitions of qualitative terms describing the performance characteristics of a measurement

Accuracy of measurement	the closeness of agreement between a quantity value obtained by measurement and the true value of the measurand
Trueness	the closeness of agreement between the average value obtained from a large series of test results and an accepted reference value
Precision	the closeness of agreement between independent test results obtained under stipulated conditions

Figure 5.2 **Relationships between type of error, qualitative performance characteristics and their quantitative expression**

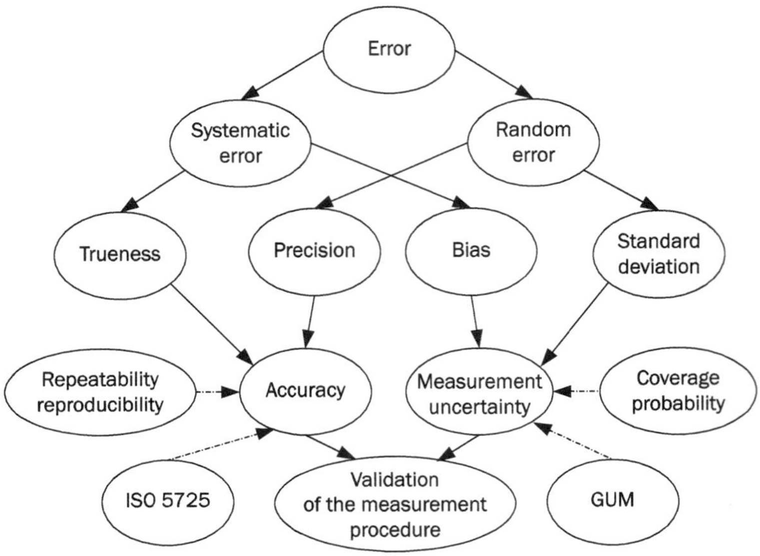

As we all know, the actual error cannot be obtained because the true value is not possible to know. Thus, the theory of uncertainty appears to evaluate the quality of measurements.

Since 1994, the International Organization of Standardization (ISO) has produced two important documents on uncertainty measurement, which are ISO 5725:1994 (Part 1–6) and ISO 'Guide to the expression of Uncertainty in Measurement' (GUM) published in 1995. Both ISO 5725 and GUM proposed a different method to express the uncertainty of a measurement. The interest of ISO in this topic relies on the consideration that quantitative evaluation of the measurement uncertainty is necessary to judge whether the differences among the measuring results of a test or the variability of a process are due to either an out-of-control process or to an incapable measurement system. This is relevant because process variation is becoming small due to technological improvements and therefore measurement procedures must be deeply tested to assess their adequateness (Pavese et al., 2009).

Many different factors (apart from variations between supposedly identical specimens) may result in the variability of measurement values from a measurement method, including:

- the operator;
- the instrument used;
- the calibration of the instrument;
- the environment (temperature, humidity, air pollution, etc.);
- the time elapsed between measurements.

The variability between measurements performed by different operators and/or with different instruments will generally be greater than the variability between measurements carried out within a short interval of time by a single operator using the same instrument.

5.3 Uncertainties in measurements

Since error is an idealized concept and errors cannot be evaluated exactly, the old concept of error disappears and the new concept of measurement uncertainty is introduced. Uncertainty is an estimate of the limits to which we can expect an error to go, under a given set of conditions as part of the measurement process. The changeover from error to uncertainty is officially set by the publication, in 1992, of the IEC-ISO 'Guide to the Expression of Uncertainty in Measurement' (GUM), which provides general rules for evaluating and expressing uncertainty for a wide range of measurements. In GUM, the definition of measurement uncertainty is 'a parameter, associated with results of a measurement, that characterizes the dispersion of the values that could reasonably be attributed to the measurand'.

Section 5.4.6 of standard ISO/IEC 17025 requires explicitly that testing laboratories should apply procedures for estimating measurement uncertainty, according to GUM and also traceability rules. Uncertainty is not error, only an estimate of its limits. The uncertainty of the result of a measurement reflects the lack of exact knowledge of the value of the measurand. The result of a measurement after correction for recognized systematic effects is still only an estimate of the value of the measurand because of the uncertainty arising from random effects and from imperfect correction of the result for systematic effects. The uncertainties can be grouped as type A and type B in a classical way. The purpose of the type A and type B classification is to indicate the two different ways of evaluating uncertainty components and is for convenience of discussion only; the classification is not meant to indicate that there is any difference in the nature of the components resulting from the two types of evaluation. Both types of evaluation are based on probability distributions and the uncertainty components resulting from either type are quantified by variances or standard deviations.

5.3.1 Type A evaluation of standard uncertainty

A type A evaluation of standard uncertainty may be based on any valid statistical method for testing data. Examples are calculating the standard deviation of the mean of a series of independent observations; using the method of least squares to fit a curve to data in order to estimate the parameters of the curve and their standard deviations; and carrying out an analysis of variance in order to identify and quantify random effects in certain kinds of measurements. If the measurement situation is especially complicated, considering the measured values x_i as the individual measurements $(x_{i1}, x_{i2}, \ldots, x_{in})$, the calculation process of type A standard uncertainty is demonstrated as equations 5.2, 5.3, 5.4 and 5.5:

$$\bar{x}_i = \frac{1}{n}\sum_{k=1}^{n} x_{ik} \tag{5.2}$$

where $\bar{x}_i$ is the mean of the measurements, and n is the number of x_i values.

$$s(x_{ik}) = \sqrt{\frac{1}{n-1}\sum_{k=1}^{n}(x_{ik} - x_i)^2} = \sqrt{\frac{1}{n-1}\sum_{k=1}^{n} v_{ik}^2} \tag{5.3}$$

where $s(x_{ik})$ is the total standard deviation of the mean.

The type A standard uncertainty component of x_i can be defined as equation (5.4):

$$u(x_i) = s(x_i) = s(x_{ik})/\sqrt{n} = \sqrt{\frac{1}{n(n-1)}\sum_{k=1}^{n}(x_{ik} - x_i)^2} \tag{5.4}$$

Considering all the uncertainty (j) components of type A, the total standard uncertainty of type A is defined as:

$$u_A = \sqrt{\sum_j (u^2(x_{ij})/n)} \tag{5.5}$$

5.3.2 Type B evaluation of standard uncertainty

A type B evaluation of standard uncertainty is usually derived from scientific judgement using all the relevant information available, which may include (Fraden, 2005):

- previous measurement data;
- experience with, or general knowledge of, the behaviour and property of relevant materials and instruments;

- manufacturer's specifications;
- data provided in calibration and other reports;
- uncertainties assigned to reference data taken from handbooks.

For the low-level calibration system, all type B uncertainties are assumed to be independent and to have rectangular or uniform distributions (that is, each error has a uniform probability of being within the region $\pm\delta_j$ and zero probability of being outside that region). If the distribution is rectangular, the standard deviation σ_s of each uncertainty component is equal to, $\delta_j/\sqrt{3}$, and the total standard deviation is, $\sqrt{(\Sigma\sigma_s^2)}$, where the summation is performed over all uncertainty components of type B.

5.4 Sources of uncertainties

In practice, there are many possible sources of uncertainty in a measurement, including '(International Organization for Standardization, 1993):

- incomplete definition of the measurand;
- imperfect realization of the definition of the measurand;
- non-representative sampling – the sample measured may not represent the defined measurand;
- inadequate knowledge of the effects of environmental conditions on the measurement or imperfect measurement of environmental conditions;
- personal bias in reading analogue instruments;
- finite instrument resolution or discrimination threshold;
- inexact values of measurement standards and reference materials;

- inexact values of constants and other parameters obtained from external sources and used in the data-reduction algorithm;
- approximations and assumptions incorporated in the measurement method and procedure;
- variations in repeated observations of the measurand under apparently identical conditions.

5.5 The evaluation of measurement uncertainties

GUM has harmonized the methods of evaluating the uncertainty of measurements using type A and type B evaluation methods. Thus, all the standard uncertainty components arising from various variables which contribute to the uncertainty can be evaluated by either the type A or type B method of uncertainty evaluation. The evaluated value of standard uncertainty is characterized by the variance or standard deviation of the random variable.

When the desired measurand Y cannot be measured directly, its value is derived from the values of N input quantities $x_1, x_2, \ldots, x_N$, modelled as random variates and functionally related to Y as:

$$Y = f(x_1, x_2, \ldots, x_N) \tag{5.6}$$

It is worth noting that the function f could model not only a physical law but also a complex measurement process and, in particular, it should contain all quantities that can contribute a significant uncertainty to the measurement result. To obtain the value of combined standard uncertainty, GUM has recommended the use of the law of propagation of uncertainty, illustrated in Figure 5.3.

Figure 5.3 Illustration of the law of propagation of uncertainty

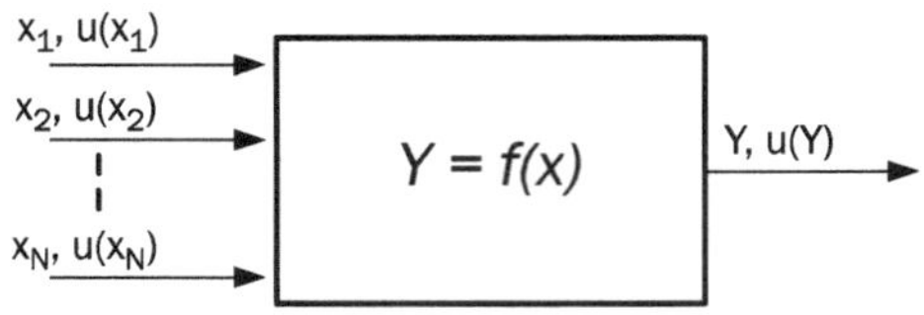

Based on a first-order Taylor series approximation, the law of propagation of uncertainty for the correlated input variables is given as follows (Kimothi, 2002; Angrisani et al., 2006):

$$u_c^2(y) = \sum_{i=1}^{N}\left[\frac{\partial f}{\partial x_i}\right]^2 u^2(x_i) + 2\sum_{i=1}^{N-1}\sum_{j=i+1}^{N}\frac{\partial f}{\partial x_i}\cdot\frac{\partial f}{\partial x_j}u(x_i,x_j) \tag{5.7}$$

where $u(x_i)$ is either the standard uncertainty on x_i or an estimate of it, and $u(x_i, x_j)$ is the estimated covariance of x_i and x_j; the value of $u(x_i, x_j)$ is given by $r(x_i, x_j) \cdot u(x_i) \cdot u(x_j)$, where $r(x_i, x_j)$ stands for the correlation coefficient associated with measured values x_i and x_j. In terms of the correlation, the law of propagation of uncertainty can also be rewritten as follows (Dunn, 2010):

$$u_c^2(y) = \sum_{i=1}^{N}\left[\frac{\partial f}{\partial x_i}\right]^2 u^2(x_i) + 2\sum_{i=1}^{N-1}\sum_{j=i+1}^{N}\frac{\partial f}{\partial x_i}\cdot\frac{\partial f}{\partial x_j}u(x_i)u(x_j)\,r(x_i,x_j) \tag{5.8}$$

where $r(x_i, x_j)$ is defined as the ratio (Kessel, 2002):

$$r(x_i,x_j) = \frac{Cov[x_i,x_j]}{\sqrt{Var[x_i]}\times\sqrt{Var[x_j]}} \tag{5.9}$$

With $Cov[xi, xj]$ being the covariance of x_i and x_j determined from the distributions. The correlation coefficients are the measure in which potential dependencies of the knowledge of the input quantities from the evaluation are expressed. Their value lies in the range: $-1 \le r(xi, xj) \le 1$.

When measurements are made under fixed conditions, the recorded values of a variable still will vary to an extent. This implicitly is caused by small variations in uncontrolled variables. The extent of these variations can be characterized by the variance of the values. Further, as the number of acquired values becomes large ($N \geq 30$), the sample variance approaches its true variance, σ^2, and the distribution evolves to a normal distribution.

Thus, if uncertainty is considered to represent the range within which an acquired value will occur, then uncertainty can be viewed as an estimate of the true variance of a normal distribution.

5.5.1 Case 1: correlated input quantities

If two or more input quantities are independent and are correlated, the combined standard uncertainty $U_c(y)$ is given by equation 5.8 (Kimothi, 2002).

5.5.2 Case 2: uncorrelated input quantities

If two or more input quantities are independent and are not correlated, the combined standard uncertainty $U_c(y)$ is the positive square root of the combined variance $U_c^2(y)$, which is obtained from the following equation:

$$u_c^2(y) = \sum_{i=1}^{N} \left[\frac{\partial f}{\partial x_i} \right]^2 u^2(x_i) \qquad (5.10)$$

In the most frequent cases, the knowledge of the input quantities can be considered non-correlated.

The expanded uncertainty U at desired level of confidence, p, is determined by multiplying the combined uncertainty by a coverage factor, for example, $k = 2$ (this specifies an interval

Figure 5.4 **Schematic of measurement uncertainty classification**

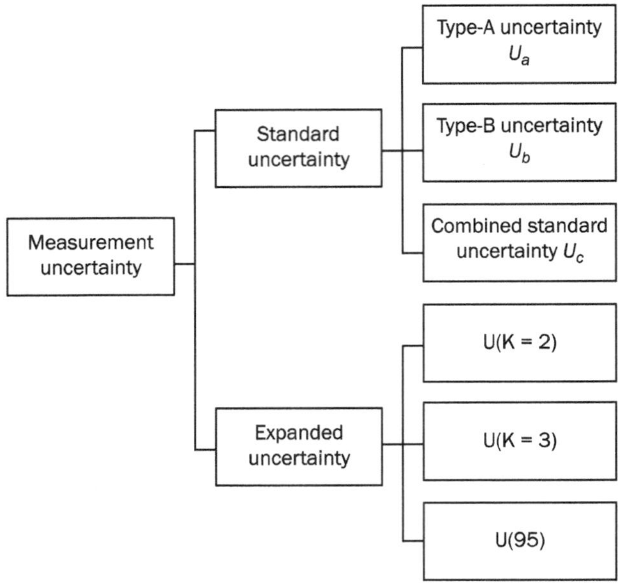

having a confidence of approximately 95 per cent). The expanded uncertainty U is then defined as:

$$U = k \cdot u_c \tag{5.11}$$

where U is an expanded uncertainty, k coverage factor chosen on the base of the desired level of confidence p to be associated with the interval U. Figure 5.4 illustrates the schematic of measurement uncertainty classification.

5.6 Conclusion

Since the ISO/IEC 17025 standard, 'General requirements for the competence of testing and calibration laboratories', was issued in 2000, the concept of measurement uncertainty on test results seems to become a key issue for laboratories. The nature of the testing method for electric motors, pumps and fans may preclude rigorous, metrologically and statistically valid calculation of measurement uncertainty. In these cases the testing laboratory shall at least attempt to identify all the components of uncertainty and make a reasonable estimation, and shall ensure that the form of reporting of the result does not give a wrong impression of the uncertainty.

6

Testing methods for electric motors

Abstract: The energy-efficiency definitions for electric motors, pumps and fans are described respectively. The IEEE and IEC testing methods standards for electric motors are discussed. The ISO testing method standard for pumps is given. The ISO and AMCA testing method standards for fans are given. The test set-ups for electric motors, pumps and fans are presented.

Key words: energy-efficiency, testing methods, standards, testing set-up.

6.1 Motor losses and efficiency expression

Motor efficiency is a measure of the effectiveness with which electrical energy is converted to mechanical energy. Motor efficiency can be directly expressed as the ratio of power output to power input (see equation 6.1). The efficiency computed from this ratio is known as the direct efficiency. Alternatively, motor efficiency is indirectly determined by using the losses and the input power (see equation 6.2):

$$\eta = \frac{P_{out}}{P_{in}} \tag{6.1}$$

$$\eta = \frac{P_{in} - P_{loss}}{P_{in}} \qquad (6.2)$$

where P_{in} and P_{out} are the input and output powers of the motor, respectively, and P_{loss} is the total losses in the motor.

Although the efficiency on 3/4 load and 1/2 load may be provided, motor efficiency is usually given for full rated load. The efficiency of a motor is affected by load, rated power and speed, as indicated below:

- A change in efficiency as a function of load is an inherent characteristic of motors. Operation of the motor at loads substantially different from rated load may result in a great change in motor efficiency, as can been seen in Figure 6.1 (in Figure 6.1, the rated load is 7.38Nm).
- Generally, the full-load efficiency of motors increases with physical size and rated output of motors.
- The efficiency of motors increases with rotation speed generally, until the peak efficiency is reached, then the efficiency decreases along with the increasing of rotation speed (see Figure 6.2).

Figure 6.1 Torque vs motor efficiency

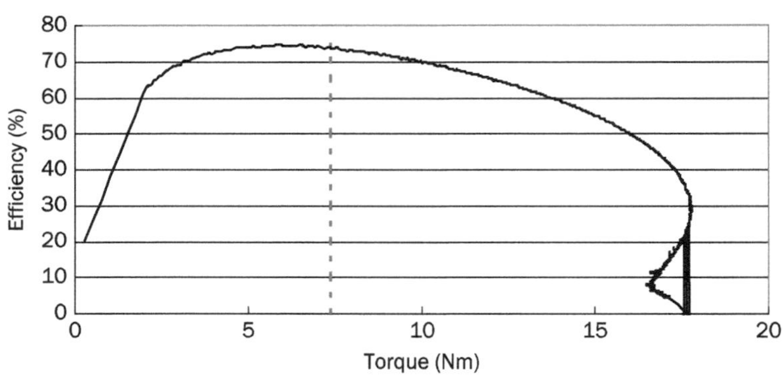

Figure 6.2 Speed vs motor efficiency

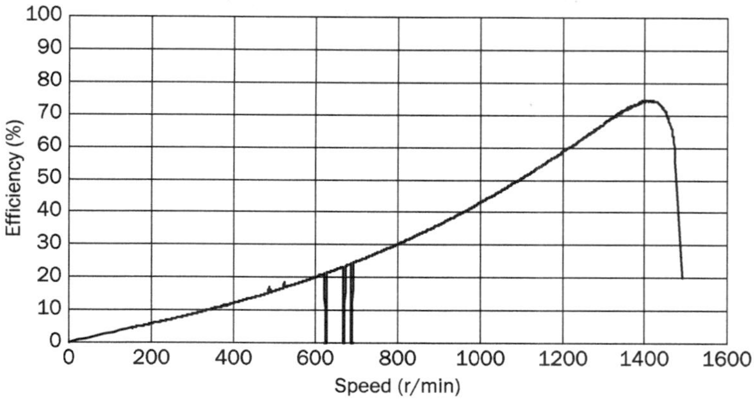

The total losses in the motor are of interest here, since calculating the efficiency is mainly related with finding the total losses. IEEE 112 Clause 5 defines five types of loss in induction motors as equation 6.3:

$$P_{loss} = P_{stator} + P_{rator} + P_{core} + P_{fw} + P_{stray} \tag{6.3}$$

where P_{stator} is stator loss, P_{rator} is rotor loss, P_{core} is core loss, P_{fw} is windage/friction loss, and P_{stray} is stray (also named additional) load loss. The first four types of loss are evaluated from measurements of input power, voltages, currents; rotational speed during testing and the remaining stray load loss is obtained based on different methods depending of the standard used. The five different loss components are described below.

Stator loss refers to the energy loss in the stator windings. It equals to $1.5I^2R$ for a three-phase motor, where I is the measured or calculated rms current per line terminal at specified load; and R is the average DC resistance between any two line terminals corrected to the specified temperature. Stator losses appear as heating due to current flow (I) passing through the resistance (R) of the stator windings. These losses are major losses and typically account for 55–60 per cent of the total losses.

Rotor loss refers to the energy loss in the rotor windings. Current flowing through the rotor windings creates rotor loss. Rotor losses appear as I^2R heating in the rotor winding. Rotor losses can be reduced by increasing the size of the conductive bars and end rings to produce lower rotor losses; another form of power loss is called slip loss because it is largely but not entirely dependent on the degree of slip the motor displays. Rotor I^2R losses are a function of the rotor conductors (usually aluminium) and the rotor slip. Slip is the difference in rpm between the rotational speed of the magnetic field and the actual rpm of the rotor and shaft at a given load.

Core loss is magnetic loss which occurs in the steel components of a motor, caused by eddy currents and hysteresis effects in the laminations and are influenced by the motor flux and frequency. It varies approximately with the square of the input voltage, but for fixed input voltage it remains approximately constant from no load to full load. It is usually measured from the no-load test. Like I^2R losses, they manifest themselves as heat. Using larger cross-sections of iron in the stator and rotor, thinner laminations and improved magnetic materials are common ways of reducing core loss.

Windage and friction loss is the mechanical rotational loss due to the friction and windage. Like core loss, it is also nearly constant from no load to full load and usually measured from the no-load test. Windage and friction losses are losses in the moving parts, such as the air gap between the rotor and the stator, friction in the bearings at each end of the rotor and any cooling fan attached to the rotor. Mechanical losses occur in the bearings, fans and seals of the motor. These losses are generally small in *IP4x* and *IP5x* slow-speed motors, but may be appreciable in large, high-speed or totally enclosed *IP6x* motors.

Stray load loss is the loss in a motor not accounted for by the sum of, P_{stator}, P_{rator}, P_{core} and P_{fw}, which is the result of leakage fluxes (I) through the resistance (R) of the stator winding and appear only when the motor is operating at under load. It can be divided into two parts: stator-stray load loss and rotor-stray load loss. Stray load losses are hard to model and to quantity. Stray load losses can be described as the losses that remain after stator and rotor losses, core losses and windage and friction losses have been accounted for. These losses vary according to square of the load current and are caused by leakage flux induced by load currents in the laminations and account for 4–5 per cent of total losses.

Typical distribution of the induction motor losses as a function of the load is shown in Figure 6.3 (Saidur, 2010). The core losses and the windage and friction losses are known as fixed or constant losses because they are considered independent of load. The other losses are considered load dependent and therefore variable with load. The conventional

Figure 6.3 **Typical distribution of the induction motor losses as a function of the load**

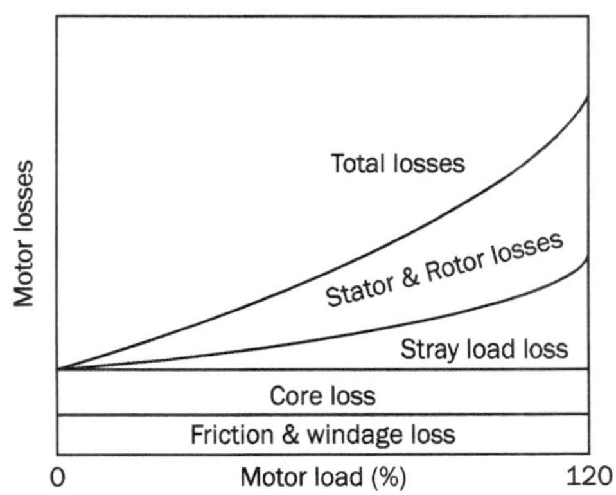

losses (stator, rotor, core and windage and friction) can be determined directly in the test process.

6.2 The status of the efficiency testing methods for IM

6.2.1 The testing methods for the efficiency of IM

Several kinds of method exist when determining the energy-efficiency of induction motors, as shown in Table 6.1 (IEEE Power Engineering Society, 2004). Generally, these methods are classified into three types: direct efficiency determination, indirect efficiency determination – sometimes called losses analysed methods – and experiment and theory method.

6.2.2 The requirements for the test power supply

Voltage

Waveform

The power supply shall provide balanced voltages closely approaching a sinusoidal waveform. The harmonic distortion coefficient, *THD*, shall not exceed 0.05. The *THD* is defined as shown in equation 6.4:

$$THD = \frac{\sqrt{E^2 - E_1^2}}{E_1} \tag{6.4}$$

where E_1 is the root-mean-square value of the fundamental of the voltage wave, in volts (V), E is the total root-mean-square value of the voltage wave, in V.

Table 6.1 **Efficiency testing methods for induction motors in IEEE 112 standard**

No.	Abbreviation name	Full name
1	Method A	Input-output
2	Method B	Input-output with segregation of losses and indirect measurement of stray-load loss
3	Method B1	Input-output with segregation of losses, indirect measurement of stray-load loss and an assumed temperature
4	Method C	Duplicate machines with segregation of losses and indirect measurement of stray-load loss
5	Method E	Electric power measurement under load with segregation of losses and direct measurement of stray-load loss
6	Method E1	Electric power measurement under load with segregation of losses and assumed value of stray-load loss
7	Method F	Equivalent circuit with direct measurement of stray-load loss
8	Method F1	Equivalent circuit with assumed value of stray-load loss
9	Method C/F	Equivalent circuit calibrated per Method C load point with indirect measurement of stray-load loss
10	Method E/F	Equivalent circuit calibrated per Method E load point with direct measurement of stray-load loss
11	Method E1/F1	Equivalent circuit calibrated per Method E load point with assumed value of stray-load loss

Voltage unbalance

The voltage unbalance shall not exceed 0.5 per cent. The per cent voltage unbalance equals 100 times the maximum voltage deviation from the average voltage divided by the average voltage.

Frequency

For general testing, the frequency shall be within ±0.5 per cent of the value required for the test being conducted, unless otherwise specified. Any departure from the specified frequency during the test directly affects the efficiency obtained with efficiency test methods A, B and B1. When these methods are used, the frequency shall be within ±0.1 per cent of the specified test value.

Rapid changes in frequency cannot be tolerated during testing because such variations affect not only the machine being tested, but also the output measuring devices. Variations in frequency during a test shall not exceed 0.33 per cent of the average frequency.

Measurement instruments

Calibrated, high-accuracy instrumentation and accessory equipment shall be used. Either analogue or digital instruments may be used in testing.

Factors affecting accuracy, particularly with non-electronic analogue instruments, are:

- loading of the signal source;
- lead calibration;
- range, condition and calibration of the instrument.

Measuring instruments shall have an accuracy class of 0.2 in accordance with IEC 60051.

Electronic instruments are generally more versatile and have much higher input impedances than non-electronic instruments. Higher input impedance reduces the need to make corrections for current drawn by the instrument. However, high input impedance instruments can be more susceptible to noise.

Instrument transformers

Instrument transformers shall have an accuracy according to IEC 60044–1 so that the errors of the instrument transformers are not greater than 0.5 per cent for general testing or not greater than 0.3 per cent for induction machines, summation of losses method, with additional load loss determination in accordance with 8.2.2.5.1.

Torque meter

The instrumentation used to measure the torque shall have an accuracy of 0.2 per cent of full scale.

Speed meter and frequency meter

The instrumentation used to measure frequency shall have an accuracy of 0.1 per cent of full scale. The speed measurement should be accurate within 0.1 per cent, or 1 revolution per minute, whichever gives the least error.

Resistance instrument selection

Calibrated high-accuracy instrumentation shall be used. Either analogue instruments (such as a Kelvin bridge) or digital instruments may be used in testing. The instruments shall bear record of calibration, within 12 months of the test, indicating limits of the error no greater than 0.2 per cent of full scale.

Methods of measuring temperatures

The temperature of various machine parts or coolant may be measured by the following:

- alcohol thermometer;
- local temperature detector;

- embedded detector;
- winding resistance.

The instrumentation used to measure winding temperature shall have an accuracy of 1°C.

As a summary, instrumentation accuracy plays a key role in determining the accuracy of the test results. Shown in Table 6.2, the instrumentation accuracies are specified in standard IEEE112 and IEC 60034–2–1. It can be seen that the IEC 60034–2–1 defines nearly the same instrumental accuracy as the IEEE counterpart, as shown in Table 6.2 (Wenping, 2009).

Table 6.2 Instrumentation accuracy and efficiency estimation (%)

Variable	IEEE 112	IEC 60034-2-1
Instrument transformer	0.5	0.3
Voltage	0.2	0.2
Current	0.2	0.2
Torque	0.2	0.2
Speed	1 rpm	1 rpm
Frequency	0.1	0.1
Resistance	0.2	0.2
Temperature	1°C	1°C
Efficiency by WCE	0.31	0.33
Efficiency by RPBE	0.17	0.18

6.3 IEEE 112-2004 B method

The standard IEEE 112–2004 is the test procedure for polyphase induction motors and generators developed by the IEEE society. IEEE 112 has been widely accepted as being a milestone in induction motor testing standards and proven to be reliable and consistent. The first version of IEEE 112 standard was issued in 1964, with revisions in 1991, 1996 and 2004. In North America, the standard IEEE 112 is used to measure the efficiency of the induction motor. The IEEE 112 consists of five basic methods to determine the efficiency, which is A, B, C, E and F. Among them, IEEE 112 B method is a well-known indirect input-output testing method based on the segregation of losses of the motor, which is widely used for determining induction motor efficiency. It is a laboratory environment testing method, designed to have a maximum of accuracy and repeatability if the instrumentation has the desired accuracy and the test procedure is followed rigorously.

In IEEE 11B method, three measuring periods are worth to mentioning here (Agamloh, 2009).

6.3.1 Thermal test

The motor is operating at rated condition, if the temperature of the motor does not change by more than 1°C over a 30-minute period, the motor reaches the steady state thermal condition. This is also known as a heat run or temperature rise test. At the end of this test, the winding resistance measurements are used to determine the rated temperature rise of the motor and a specified temperature at which efficiency is determined. This specified temperature is defined in three ways in the IEEE 112 standard. Preferably, the specified temperature is the measured temperature rise by resistance from a rated load temperature test plus 25°C. The use of a reference ambient temperature of 25°C ensures that motors tested under different ambient conditions can be easily compared.

6.3.2 No-load test

The motor is supplied with the rated voltage and frequency, running without mechanical load until the bearings are stabilized between two consecutive measures spaced out of 30 minutes; the input power does not increase over 3 per cent, then a variable voltage test is performed.

The friction, windage and core losses are determined from a no-load test. The test is performed with the voltage varying from a high value to the lowest possible voltage. The maximum voltage is above the nominal value and the lowest possible voltage is that which any further reduction of voltage leads to an increase in current. The resistance is measured so that the resistance drop across the windings can be subtracted.

At rated voltage, the core, friction and windage losses can be obtained from equation 6.5:

$$P_0 - 3I_0^2R_0 = P_{core} + P_{fw} \quad (6.5)$$

where P_0 is the input power at no load, I_0 is the no-load current, R_0 is the stator resistance of each phase measured after allowing for what is known as bearing loss stabilization.

This is considered to have been achieved if two no-load power measurements taken 30 minutes apart do not vary by more than 3 per cent. To determine the friction and windage loss, subtract the stator *I2R* loss (at the temperature of the test) from the total losses (i.e., input power) at each of the test voltage points and plot the resulting power curve versus voltage, extending the curve to zero voltage. The P_{fw} is the intercept with zero voltage axis of a plot of the power versus voltage squared curve (see Figure 6.4).

The core loss P_{core} at each test voltage is obtained by subtracting the determined P_{fw} from the input power minus stator loss. A plot of core loss versus voltage can be

Figure 6.4 A plot of the power vs voltage squared curve

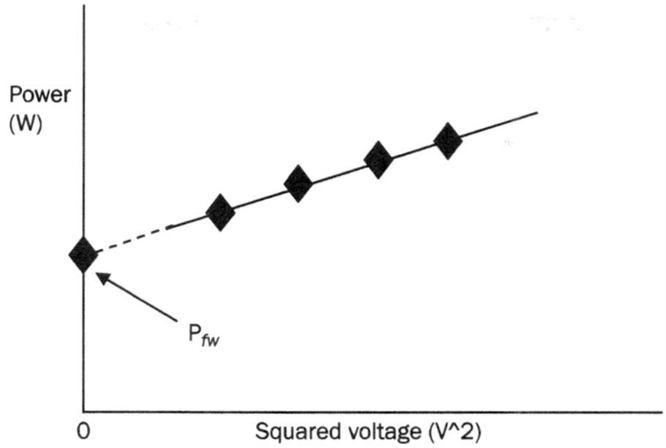

constructed for use in determining the core loss at any desired voltage. The value of P_{core} and P_{fw} are considered constant at each load in the IEEE 112B method.

6.3.3 Variable load test

After the steady thermal state and resistance measurements, the motor is restarted quickly (the motor is still hot, the winding temperature has not changed more than 10°C with reference to the rated one) and loaded with six decreasing load torque values (at 125 per cent, 110 per cent, 100 per cent, 75 per cent, 50 per cent and 25 per cent of the rated motor torque). The test data measured and recorded should include the speed, torque, input power, voltage and current, winding and ambient temperature. The stator loss is determined for each load point as the ohmic loss in the stator winding at that load:

$$P_{stator} = 1.5 I_1^2 R_1 \frac{K + \theta_W - \theta_c + 25}{K + \theta_w} = 3 I_1^2 R \frac{K + \theta_W - \theta_c + 25}{K + \theta_w} \quad (6.6)$$

where I_1 is the load current, R_1 is the DC resistance between any two line terminals corrected to the appropriate temperature, if required, R is the average stator resistance per phase, θ_w is the winding test temperature, θ_c is the inlet coolant temperature during test. The temperature constant 235 is for copper; this should be replaced by 225 for an aluminium winding.

The rotor loss is determined as a product of the slip and the air gap power. The air gap power is obtained by subtracting the core and stator losses from the power input measured at the given load point:

$$P_{rotor} = s*(P_1 - 3I_1^2R - P_{core})\frac{K+\theta_w - \theta_c + 25}{K+\theta_w} \tag{6.7}$$

where the last quantity, $P_1 - 3I_1{}^2R - P_{core}$, is the power across the air gap of the motor, P_1 is the power input measured at the load point, s is the motor slip calculated from the synchronous speed n_s and the shaft speed n. The first four losses in equation 6.3 are so far obtained from equations 6.5, 6.6 and 6.7. The stray loss is determined by subtracting the sum of the previously determined losses (stator and rotor loss, core loss, and friction and windage loss) from the apparent total loss. The apparent total loss is the difference between the input power and output power:

$$P_{str} = P_{app} - P_{core} - P_{fw} - P_{stator} - P_{rator} \tag{6.8}$$

$$P_{app} = P_{in} - P_{out} \tag{6.9}$$

where P_{str} is the raw unsmoothed stray load loss, the raw unsmoothed stray load loss may have bias with the true value because of the unstable installation of the test equipment, the poor coaxiality or measurement uncertainty. The raw stray load loss measurements are smoothed through a regression method to obtain P_{stray}. The regression analysis is done by plotting P_{str} versus torque squared, and a linear regression is used to reduce the effect of random errors in the test measurements:

$$P_{str} = A \times T^2 + B \tag{6.10}$$

where T is the machine torque, A and B are constants determined from at least six load points using the following equations:

$$A = \frac{i \times \sum((P_L) \times (T^2)) - \sum P_{str} \times \sum T^2}{i \times \sum T^4 - \left(\sum T^2\right)^2} \tag{6.11}$$

$$B = \frac{\sum P_L}{i} - A \times \frac{\sum T^2}{i} \tag{6.12}$$

where i is the number of load points summed.

The following expression is used to calculate the correlation coefficient:

$$\gamma = \frac{i \times \sum(P_L \times T^2) - (\sum P_L) \times (\sum T^2)}{\sqrt{(i \times \sum(T^4) - (\sum T^2)^2) \times (i \times \sum P_L^2 - (\sum P_L)^2)}} \tag{6.13}$$

When the correlation coefficient γ is less than 0.95, delete the worst point and repeat the regression. If γ increases to ≥ 0.95, use $P_{stray} = AT^2$ to calculate the smoothed stray load losses of each load point. The smoothed stray-load loss data are used to calculate the final value of total loss and the efficiency.

6.4 IEC 60034-2-1:2007

In IEC 60034–2–1:2007, the stray load losses are no longer assumed to be a lump sum value of 0.5 per cent; instead, they are determined by making measurements. Previously, the

International Electrotechnical Commission (IEC) standard 34–2 was widely used in Europe for determining induction motor efficiency. The new standard, IEC 60034–2–1:2007, presents a kind of energy-efficiency testing method, which is indirect loss determination. It is a convenient testing method; however, it has some shortages which are not so important when testing low-efficiency motors but which have a great influence on the testing results of high-efficiency motors. The record form is illustrated in Appendix A (© [2004] IEEE) at the end of this chapter.

6.5 Eh-star test

6.5.1 Test circuit for Eh-star method

The Eh-star testing method was proposed in 1967. The abbreviation 'Eh' has its origins from the German 'einphasig mit hilfswiderstand', which means 'single phase with auxiliary resistance'. This test requires operating the uncoupled motor with unbalanced voltage supply. The test circuit is shown in Figure 6.5.

The Eh-star circuit is an asymmetric feeding of a three-phase induction machine (squirrel cage or wound rotor) without coupling of the machine and without needing any dynamometer. The asymmetrical operation is obtained by operating the stator winding in the star connection (if motors that connected in delta connection are reconnected to star connection during this test.). The star point must not be connected to system neutral or earth to avoid zero-sequence currents.

Two machine phases U and W are connected in parallel through an ohmic resistance R_{eh} of value similar to the motor short-circuit impedance.

Figure 6.5 **Eh-star circuit**

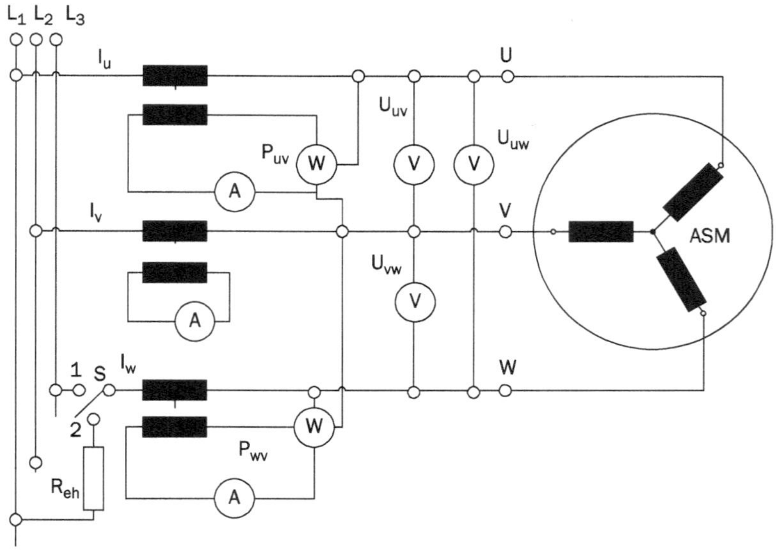

The third motor-phase shall be connected to the power line by means of a resistor R_{eh} having approximately the following typical value:

- for motors rated for star connection:

$$R'_{eh} = \frac{U_N}{\sqrt{3} \cdot I_N} \cdot 0.2 \tag{6.14}$$

- for motors rated for delta connection:

$$R'_{eh} = \frac{\sqrt{3} \cdot U_N}{I_N} \cdot 0.2 \tag{6.15}$$

The measurement instruments include three current instruments, three voltage instruments and two power instruments (two meters are needed; three-phase power instruments are forbidden to be used). The instruments shall have an accuracy of 0.2 per cent of full scale.

The resistor R_{eh} used during the test shall be adjusted so that the positive sequence current *I(1)* stays below 30 per cent of negative sequence current *I(2)* and the speed stays in the range of typical motor speeds near rated speed.

Test current I_t and test voltage U_t is given by:

- for motors rated for star connection:

$$I_t = \sqrt{I_N^2 - I_0^2};\ U_t = U_N \tag{6.16}$$

- for motors rated for delta connection:

$$I_t = \frac{\sqrt{I_N^2 - I_0^2}}{\sqrt{3}};\ U_t = \sqrt{3} \cdot U_N \tag{6.17}$$

6.5.2 *Test procedure for Eh-star method*

The preparatory work prior to the test

Prior to the test, the no-load losses have to be stabilized through tests. To avoid excessive unequal heating of the three phases, the test shall be conducted on a cold machine and carried out as quickly as possible.

In addition, the resistance between the terminals *V* and *W* (R_{VW}) shall be measured and recorded before and after the complete test. The winding temperature θ_1 shall be recorded if the resistance is obtained according to the temperature variation.

Test procedure and notes

1. During the test

During the test, the resistance R_{eh} shall be adjusted so that the positive sequence current I_{ps} stays below the 30 per cent of negative sequence current I_{ne}, and the speed stays in the range of typically rated speed (Aoulkadi and Binder, 2006). At the start of the test, the difference between the actual resistor R_{eh} and calculated value R'_{eh} should not be bigger than 20 per cent.

2. Test current

- For motors rated for star connection:

$$I_t = \sqrt{I_N^2 - I_0^2} \tag{6.18}$$

- For motors rated for delta connection:

$$I_t = \sqrt{I_N^2 - I_0^2} / \sqrt{3} \tag{6.19}$$

3. Test voltage

- For motors rated for star connection:

$$U_t = U_N \tag{6.20}$$

- For motors rated for delta connection:

$$U_t = \sqrt{6} U_N \tag{6.21}$$

4. Start-up

Larger motors can only be started without the R_{eh} resistor at reduced voltage (25–40 per cent U_N). After run-up, connect R_{eh} by switching to position 2. Smaller motors should start up with resistor R_{eh} already connected.

5. Test points

Vary the supply voltage for six test points. The test points shall be chosen to be approximately equally spaced between 150 per cent and 75 per cent of rated phase current measured in phase $V(I_V)$. When starting the test, begin with the highest current and proceed in descending order to the lowest current. Record for each test point: (I_U, I_V, I_W); (U_{UV}, U_{VW}, U_{WU}); (P_{UV}, P_{WV}), n.

6. *Winding temperature*

If the resistor of the winding is calculated according to the temperature of the winding, which is obtained by the temperature measuring sensor setting on the winding, at each test point the winding temperature θ_2 is measured and recorded. Otherwise, after the test, the resistor R_{vw} shall be recorded, and afterwards five points should be recorded continuously.

7. *Obtaining the resistor R_{vw}*

The resistor R_{vw} between phase V and W is obtained according to the following rules:

- When the resistor value R_{vw} is obtained by direct measurement, the test resistance at the end of the thermal test shall be determined in a way similar to the extrapolation procedure as described in 8.6.2.3.3 of IEC 60034–1, using the shortest possible time instead of the time interval specified in Table 5 therein, and extrapolating to zero.
- When winding resistance (under load) cannot be measured directly, the test resistance value R_2 shall be calculated by the resistor R_1, temperature θ_1 which is measured prior to the test, and the temperature θ_2, which is measured at each test point:

$$R_2 = \frac{R_1(235+\theta_2)}{(235+\theta_1)} \tag{6.22}$$

8. Achieving accurate results

To achieve accurate results, the slip shall be not greater than twice the rated slip for all currents; in other words: $n > n_{syn} - 2(n_{syn} - n_N)$. If this condition cannot be met, the test shall be repeated with an increased value of R_{eh}. If the motor still runs unstable at currents below 100 per cent of rated phase current, these test points should be omitted.

The line-to-line resistance R_{VW} for 100 per cent test current and lower currents shall be the value determined after the lowest reading (at the end of the test). The resistance used for currents higher than 100 per cent shall be determined as being a linear function of current, using the readings before and after the complete test. The test resistance is determined using the extrapolation according to item 5.7.1 in IEC 60034–2–1.

6.5.3 The stray losses computation in the Eh-star test (IEC, 2007)

1. Determine the following complex voltages and currents from the test results:

$$\underline{U}_{uv} = U_{uv} \tag{6.23}$$

$$U_{wu} = \frac{U_{vw}^2 - U_{wu}^2 - U_{uv}^2}{2U_{uv}} \tag{6.24}$$

$$U''_{wu} = \sqrt{U_{wn}^2 - U'^2_{wu}} \tag{6.25}$$

$$U'_{vw} = -\underline{U}_{uv} - U'_{wu} \tag{6.26}$$

$$U'_{vw} = -U''_{wu} \tag{6.27}$$

$$I'_v = -\frac{(P_{uv} - P_{vw}) + U_{wu} I_w}{U_{uv}} \tag{6.28}$$

$$I''_v = \sqrt{I_v^2 - I_v'^2} \tag{6.29}$$

$$k_1 = \frac{1}{2 \cdot I_v^2} \cdot \left(I_w^2 - I_u^2 - I_v^2\right) \tag{6.30}$$

$$I'_u = K_1 \cdot I'_v + \sqrt{(k_1^2 - \frac{I_u^2}{I_v^2})(I_v'^2 - I_v^2)} \tag{6.31}$$

$$I''_u = \frac{k_1 I_v^2 - I'_u \cdot I'_v}{I''_v} \tag{6.32}$$

$$I'_w = -I'_u - I'_v \tag{6.33}$$

$$I'_w = -I''_u - I''_v \tag{6.34}$$

2. Determine the inner line-to-line voltages from the complex line-to-line voltages and currents:

$$\underline{U}_{iUV} = \underline{U}_{UV} + \frac{R_{VW}}{2} \cdot \left(\underline{I}_V - \underline{I}_U\right) \tag{6.35}$$

$$\underline{U}_{iVW} = \underline{U}_{VW} + \frac{R_{VW}}{2} \cdot \left(\underline{I}_W - \underline{I}_V\right) \tag{6.36}$$

3. Separate into positive and negative sequence line-to-line components ($\underline{a} = e^{j2\pi/3}$):

$$\underline{U}_{iLL(1)} = (1)\frac{1}{3} \cdot (\underline{U}_{iUV} + \underline{a} \cdot \underline{U}_{iVW} + \underline{a}^2 \cdot \underline{U}_{iWU}) \tag{6.37}$$

$$\underline{U}_{iLL(2)} = \frac{1}{3} \cdot (\underline{U}_{iUV} + \underline{a}^2 \cdot \underline{U}_{iVW} + \underline{a} \cdot \underline{U}_{iWU}) \tag{6.38}$$

4. Determine the positive and negative sequence components of the inner-phase voltage $\underline{U}_i$:

$$\underline{U}_{i(1)} = \frac{1}{\sqrt{3}} \cdot e^{-j\frac{\pi}{6}} \cdot \underline{U}_{iLL(1)} \tag{6.39}$$

$$\underline{U}_{i(2)} = \frac{1}{\sqrt{3}} \cdot e^{-j\frac{\pi}{6}} \cdot \underline{U}_{iLL(2)} \tag{6.40}$$

5. Determine the asymmetrical inner-phase voltages:

$$\underline{U}_{iU} = \underline{U}_{i(1)} + \underline{U}_{i(2)} \tag{6.41}$$

$$\underline{U}_{iV} = \underline{a}^2 \cdot \underline{U}_{i(1)} + a \cdot \underline{U}_{i(2)} \tag{6.42}$$

$$\underline{U}_{iW} = \underline{a} \cdot \underline{U}_{i(1)} + a^2 \cdot \underline{U}_{i(2)} \tag{6.43}$$

6. Determine the iron loss resistance:

$$R_{fe} = \frac{U_t^2}{P_{fe}} \tag{6.44}$$

7. Determine the inner-phase currents:

$$\underline{I}_{iU} = \underline{I}_U - \underline{I}_{feU} \tag{6.45}$$

$$\underline{I}_{iV} = \underline{I}_V - \underline{I}_{feV} \tag{6.46}$$

$$\underline{I}_{iW} = \underline{I}_W - \underline{I}_{feW} \tag{6.47}$$

8. Determine the positive and negative sequence components of the inner-phase currents:

$$\underline{I}_{i(1)} = \frac{1}{3} \cdot (\underline{I}_{iU} + \underline{a} \cdot \underline{I}_{iV} + \underline{a}^2 \cdot \underline{I}_{iW}) \tag{6.48}$$

$$\underline{I}_{i(2)} = \frac{1}{3} \cdot (\underline{I}_{iU} + \underline{a}^2 \cdot \underline{I}_{iV} + \underline{a} \cdot \underline{I}_{iW}) \tag{6.49}$$

9. Determine the air gap power:

$$P_{\delta 1} = 3 \cdot (U'_{i(1)} \cdot I'_{i(1)} + U''_{i(1)} \cdot I''_{i(1)}) \tag{6.50}$$

$$P_{\delta 2} = 3 \cdot (U'_{i(2)} \cdot I'_{i(2)} + U''_{i(2)} \cdot I''_{i(2)}) \tag{6.51}$$

10. Determine the additional load losses:

$$P_{Lr} = k \cdot [(1 - s) \cdot (P_{\delta 1} - P_{\delta 2}) - P_{fw}] \tag{6.52}$$

Where $k = \frac{1}{1 + (I_{i(1)} / I_{i(2)})^2}$.

6.6 GB/T 9651

Standard GB/T 9651 is the test procedure for the single-phase induction motor. In this standard, the efficiency for the single-phase induction motor is determined by a direct measurement method. The input power of the motor is measured by a wattmeter and the output power is measured by the following methods (Standardization Administration of China, 2008):

- dynamometer method;
- rotor torque measurement method;
- slode and rod method.

The dynamometer is connected with the test motor by coupler; the rotor torque meter is connected with the test motor according to the specified usage requirement.

During the test, the test motor shall reach the heat stability condition. Among the 1.25–0.25 of the rated power, the work characteristic curve is determined when the load is descended and increased.

For the small power motor, the characteristic curve can be drawn only according to the descended loads. Six to eight points are measured for each curve; at each measuring point, the following values shall be obtained: the stator voltage, stator current, input power, output torque, speed and the resistor of stator winding, and the environment temperature.

6.7 Energy-efficiency test rig for electric motors

Both the direct and indirect efficiency testing methods involve loading the motor with a dynamometer and measuring the speed, torque, voltage, current, power factor and power input. Generally, three types of dynamometers exist, each with a different braking system to absorb load: hysteresis, eddy-current and powder.

Hysteresis brake dynamometers are versatile and ideal for testing in the low to middle power range, as shown in Figure 6.6. Eddy-current brake dynamometers are ideal for applications requiring high speeds and also when operating in the middle to high power range. Powder brake dynamometers are ideal for applications operating in the low to middle speed range or when operating in the middle to high torque range.

The test set-up required for determining the motor's efficiency based on the summation of losses method depends mainly on how stray-load losses are evaluated. If these losses are computed as the residual losses, a torque measurement is

Figure 6.6 Hysteresis brake dynamometers

required, meaning the use of a calibrated dynamometer with a load cell or an in-line torque meter. For all standards mentioned, this is the preferred method. Stray-load losses may also be evaluated by a reverse rotation test, an assigned value or by an Eh-star test. In these cases, no torque measurement is required but uncertainty in the efficiency should be higher than with the residual losses approach.

A schematic of a motor performance test rig is shown in Figure 6.7. Readings of electrical power, current, voltage, frequency, speed or slip, torque, stator winding temperature or stator winding resistance, and ambient temperature shall be obtained at each load point. In loading the machine, start at the highest load value and move in descending order to the lowest. Test procedure of IEEE 112 method B for the efficiency of electric motors is listed in Table 6.3.

Figure 6.7 Schematic of a motor test rig

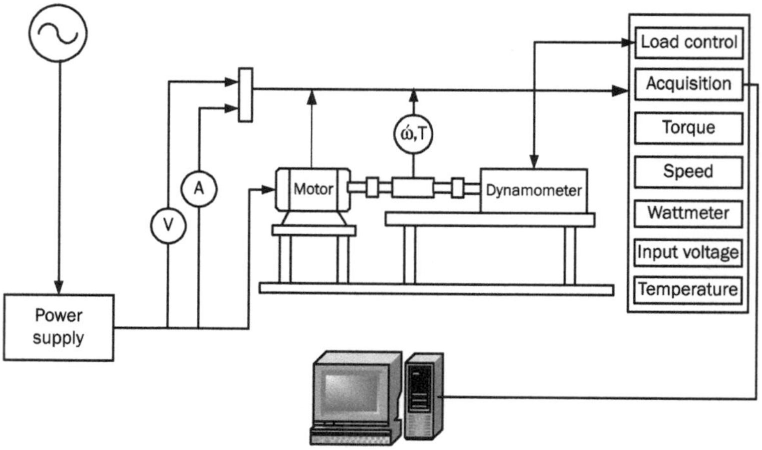

Table 6.3 **Test procedure for the efficiency of electric motors**

Main steps	Measuring contents
1	Installation of temperature detector on the stator winding/core/frame
2	Measurement of motor cold resistance and temperature
3	Temperature test at rated load
4	Measurement of motor-related load resistance and temperature
5	Measurement of motor resistance before the highest and after the lowest load point
6	Load test
7	No-load stabilization of the input power
8	No-load test

6.8 Conclusion

In this chapter, the energy-efficiency expressions for electric motors have been provided. The measuring parameters which determine the energy-efficiency for electric motors have been pointed out. Direct and indirect energy-efficiency testing methods for the electric motor have been shown. Among these, the IEEE 112B method has been analysed in detail because it is regarded as the 'golden' method for its precision and repeatability. The energy-efficiency test rigs for the electric motor have been demonstrated and discussed.

6.9 Appendix A (Used with permission of IEEE Power Engineering Society)

Type: **Design:** **Frame:** **hp:**
Freq: **Volts:** **Synch. r/min:** **Serial No:**
Deg C Temp Rise: **Model No:** **Phase:**

Stator Winding Resistance Between Terminals @ deg C							
Specified Temperature for Resistance Correction = deg C							
Item	Load, in per cent rated	1	2	3	4	5	6
1	Ambient Temperature, in C						
2	Stator Winding Temperature, in C						
3	Slip, in r/min						
4	Speed, in r/min						
5	Line-to-Line Voltage, in V						
6	Line current, in A						
7	Stator power, in W						
8	Core Loss, in W						
9	Stator I^2*R loss, in W, at(tt)C						
10	Power across air gap, in W						
11	Rotor I^2*R loss, in W						
12	Friction and windage loss, in W						
13	Total conventional loss, in W						
14	Torque, in Nm						
15	Dynamometer Correction, in Nm						
16	Corrected torque, in Nm						
17	Shaft power, in W						
18	Apparent total loss, in W						
19	Stray-load loss, in W						
Int. Slope Corr. Factor Del. Pt.							

20	Stator I^2*R Loss, in W,						
21	Corrected Power Across Air Gap, in W						
22	Corrected slip, in r/min						
23	Corrected speed, in r/min						
24	Rotor I^2*R Loss, in W						
25	Corrected stray-load loss, in W						
26	Corrected total loss, in W						
27	Corrected shaft power, in W						
28	Shaft power, in hp						
29	Efficiency, in per cent						
30	Power Factor, in per cent						

Summary of Characteristics

Load, in per cent of rated							
Power Factor, in per cent							
Efficiency, in per cent							
Speed, in r/min							
Line Current, in A							

7

Efficiency testing methods for centrifugal pumps

Abstract: The calculation formula for the efficiency of centrifugal pumps is presented. The influence variables on the efficiency of centrifugal pumps are introduced. The international testing standard for pump performance – ISO 9906 – is given. Finally, the testing rig for pump performance is introduced.

Key words: efficiency formula, pump, testing standard, testing rig.

7.1 Efficiency of pumps

Centrifugal pump performances are tested by using clean cold water or viscous oil. Pump efficiency is determined by principally two parameters, head and flow rate, in addition to other factors such as properties of the fluid, impeller design and motor speed selected.

Pumps are driven by an electric motor; the pump input power can be calculated from:

$$P_{in} = T\omega \tag{7.1}$$

where T is the torque, Nm; ω is the angular velocity, rad/s.

The power output of pumps is mechanical power transferred to the liquid during its passage through the pump. It can be calculated by:

$$P_{out} = \rho g Q H \tag{7.2}$$

where Q is the flow rate, m^3/s; H is the pump head, *m*; and ρ is the fluid density, kg/m^3; g is acceleration due to gravity, m/s^2.

Pump efficiency is expressed as pump power output divided by the pump power input, so the overall pump efficiency can be finally computed from the relationship:

$$\eta = \frac{\rho g Q H}{T\omega} \times 100\% \tag{7.3}$$

The pump head is a term derived from the Bernoulli formulation, an energy balance around the pump. It is expressed as:

$$H = \frac{(P_2 - P_1)}{\rho g} + \frac{(V_2^2 - V_1^2)}{2g} + (Z_2 - Z_1) \tag{7.4}$$

where the subscript 2 represents the value at the outlet of the pump and the subscript *1* is the value at the inlet of the pump. P is the pressure, V is the velocity, and Z is the height. In this equation, the first term '$P_2 - P_1/\rho g$' is the pressure head, the second term '$V_2^2 - V_1^2/2g$' is the velocity or kinetic head, and the third term '$Z_2 - Z_1$' is the height difference. If the diameter of the outlet pipe is equal to the diameter at the inlet of the pump, then $V_2 = V_1$. Also, if the pump is sufficiently small and the height difference is small, equation 7.4 can be defined by:

$$H\rho = P_2 - P_1 \tag{7.5}$$

Substituting equation 7.5 into equation 7.2 gives:

$$P_{out} = Q^* \ (P_2 - P_1) \tag{7.6}$$

The schematic of performance curves of pumps is shown in Figure 7.1. It is equally important to select a pump whose best efficiency point (BEP) falls as close as possible between the specified rated flow (maximum demand flow rate) and the normal flow (Figure 7.1). This will help to maximize pumping efficiency (Shiels, 1998).

Figure 7.1 **Schematic of performance curves of pump**

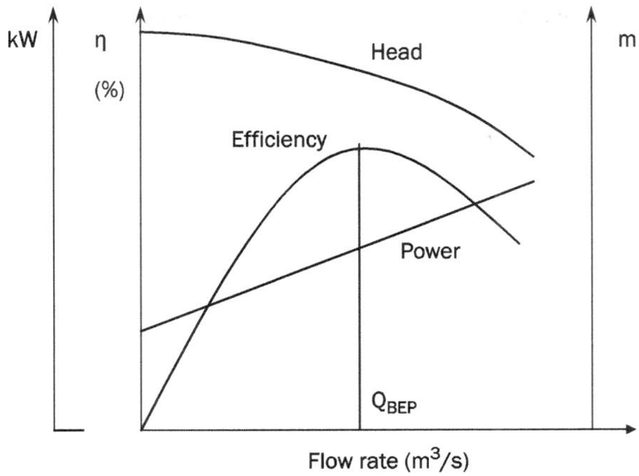

All the parameters such as flow rate, total head and input power will change if the speed is varied. It is important to have an appreciation of how these parameters vary in order to safely control a pump at different speeds. The relationships of rotodynamic pump performance parameters of flow, head and power absorbed to speed are known as the affinity laws. The equations that explain these are known as affinity laws. Flow rate (Q) is proportional to rotational speed (N), $Q \propto N$. Head (H) is proportional to the square of the rotational speed, $H \propto N^2$. Power absorbed (P) is proportional to the cube of the rotational speed, $P \propto N^3$.

7.2 ISO 9906:1999

ISO 9906:1999 'Rotodynamic pumps – Hydraulic performance acceptance tests – Grades 1 and 2' contains two grades of accuracy of measurement: grade 1 for higher accuracy and grade 2 for lower accuracy (International Standardization Organization, 1999). These grades include different values for tolerance factors, for allowable fluctuations and uncertainties of measurements. Pump inlet and exit pressures are measured by two pressure transducers. They have an accuracy of 0.5 per cent. The actual places of the pressure transducers are selected according to ISO 2548. Pumps are performance tested in accordance with ISO 9906 or corresponding test standards.

7.3 GB/T 3216–2005

GB/T 3216–2005 'Rotodynamic pumps – hydraulic performance acceptance tests – grades 1 and 2' is a China standard modified from ISO 9906:1999 (Standardization Administration of China, 2005).

7.4 Testing rig for pumps

The pump's head, flow rate, current and power consumption are measured and recorded by the testing software at rated voltage and frequency. Pumps can be tested at all common voltages and frequencies. The testing results are evaluated and output as the characteristic curves of the pump. The schematic of a pump test apparatus is shown in Figure 7.2. For each quantity to be measured, Table 7.1 gives the permissible amplitude of fluctuations of the measured parameters.

7.5 Conclusion

In this chapter, the energy-efficiency expressions for pumps have been provided. The measuring parameters which

Figure 7.2 **Schematic of a pump test apparatus**

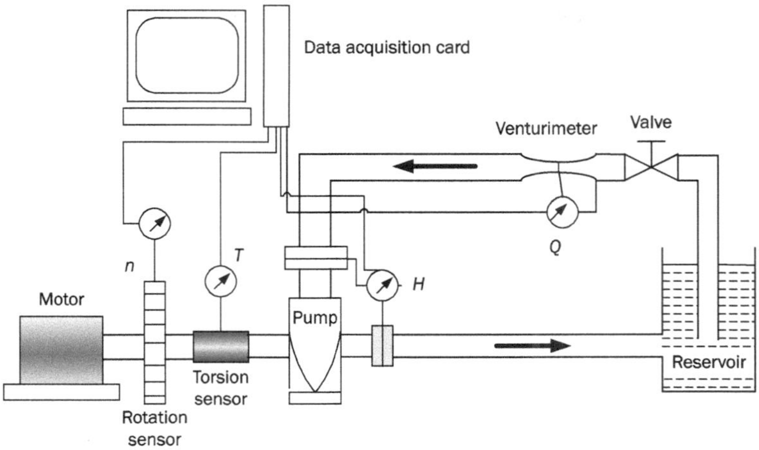

Table 7.1 **Permissible amplitude of fluctuations as a percentage of mean value of quantity being measured**

Measured quantity	Permissible amplitude of fluctuations	
	Grade 1 %	Grade 2 %
Flow rate	±3	±6
Pump total head		
Torque		
Power input		
Speed of rotation	±1	±2

When using a differential pressure device to measure flow rate, the permissible amplitude of the fluctuations of the observed differential pressure shall be ±6 per cent for grade 1 and ±2 per cent for grade 2.

In the case of separate measurements of inlet total pressure and outlet total pressure, the maximum permissible amplitude of fluctuation shall be calculated on the pump total head.

determine the energy-efficiency for pumps have been pointed out. The key measuring parameters in the energy-efficiency testing for pumps are flow rate, head and input power. In the same principle, the volume flow rate, fan power and the static pressure rise are the key variables determining the energy-efficiency of fans. The energy-efficiency test rigs for pumps have been presented and discussed.

8

Efficiency testing methods for centrifugal fans

Abstract: The calculation formula for the efficiency of centrifugal fans is given. The efficiency of centrifugal fans is related to flow rate, static pressure rise, input fan power, hub-to-tip ratio and density of the air. The international testing standards for fan performance are introduced. Finally, the testing rig for fan performance is interpreted.

Key words: efficiency formula, fans, testing standards, testing rig.

8.1 Efficiency of fans

Fan performance is typically defined by a plot of developed pressure and power required over a range of fan-generated airflow. Understanding this relationship is essential to designing, sourcing and operating a fan system and is the key to optimum fan selection.

The best efficiency point (BEP) is an important point reflected on the fan performance curve, where a fan operates most cost-effectively in terms of both energy-efficiency and maintenance considerations. Operating a fan near its BEP improves its performance and reduces wear, allowing longer intervals between repairs. Moving a fan's operating point away from its BEP increases bearing loads and noise.

Each fan has a performance curve and associated data; this details its air volume flow rate versus pressure characteristic:

$$\eta = \frac{K \Delta P_s Q}{\wp} \times 100\% \tag{8.1}$$

where:

- ΔPs is the static pressure rise;
- ρ is the density of the air;
- Q is the volume flow rate;
- K is a constant depending on the hub-to-tip ratio;
- $\wp$ is the fan power.

The roots of today's practice are based on using fan static pressure for fan selections. It is certain that the most important quantity sought has been the flow of air through a fan. Later, another quantity significant for fan selection was added – fan pressure – and that undoubtedly was the pressure we today call fan static pressure. It can be safely said that in the last five decades it has been a common practice to select fans for applications using static pressure rather than total pressure, despite the fact that the pressure required by a system must be satisfied by the fan total, not static, pressure. As a result, static efficiency has been used as the measure of efficiency because it conveniently tied the fan power to the commonly used fan pressure. That is not to say that the fan total pressure has not been used at all; however, the usage was limited primarily to selection of axial flow fans.

8.2 ISO 5801:2008

ISO 5801:2008 'Industrial fans – performance testing using standardized airways' goes some way in recognizing that there are four installation categories: type A, B, C and D (the

Figure 8.1 Fan installation categories

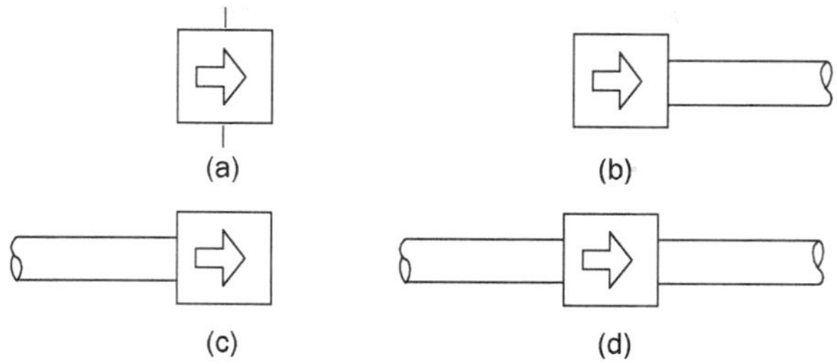

different types of configuration are shown in Figure 8.1) (International Standardization Organization, 2008). The fan to be tested shall be classified according to one of the four categories. The supplier should state the category of installation for which the fan is intended, and the user should select from the categories available the one which is closest to their application. In Figure 8.2, category A simulates the fan with free inlet and free outlet condition, category B simulates the fan with free inlet and ducted outlet condition, category C simulates the fan with ducted inlet and free outlet condition, and category D simulates the fan with ducted inlet and ducted outlet condition.

8.3 AMCA 210-2007

The US approach to energy classification for fans is designed by large professional bodies, such as AMCA International (Air Movement and Control Association International) and ASHRAE (American Society of Heating, Refrigerating, and Air-Conditioning Engineers). AMCA Standard 210–2007 'Laboratory methods of testing fans for certified aerodynamic performance ratings' is published by the AMCA International and the ASHRAE.

Figure 8.2 Fan test chamber

This standard establishes uniform methods for laboratory tests of a fan or other air-moving device to determine its aerodynamic performance in terms of airflow rate, pressure developed, power consumption, air density, speed of rotation and efficiency for rating or guarantee purposes.

8.4 GB/T 1236-2000

Chinese National Standard GB/T 1236–2000 'Test methods of aerodynamic performance for fans' is equivalent to ISO 5801:1997E.

8.5 Testing rig for fan performance

The test chamber and instrument set-up according to the AMCA standard are plotted schematically in Figure 8.3. The fan discharge is mounted on the wall of an AMCA test chamber to simulate the free-inlet and free-outlet condition. The facility is a rectangular chamber which comprises

Figure 8.3 **Schematic of a fan test facility**

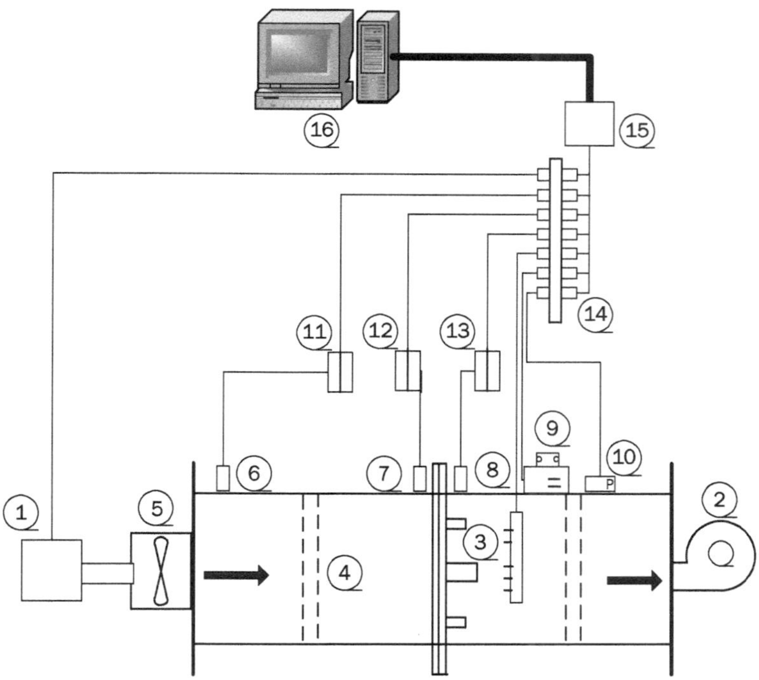

Notes: 1. optical tachnometer; 2. auxiliary fan; 3. multiple nozzles; 4. flow-settling means; 5. test fan; 6. pressure taps for measuring static pressure; 7. and 8. pressure taps for measuring flow rate; 9. psychrometer; 10. pitot tube; 11. press trans no. 1; 12. press trans no. 2; 13. press trans no. 3; 14. data acquisition; 15. sensor input card; 16. computer; 17. thermometer.

flow-settling means, pressure measuring devices, nozzles and an auxiliary fan. Flow-settling means are installed in the chamber to provide proper flow patterns for measurements (Lin and Huang, 2002). For a measuring plane located upstream of the settling means, the settling screen can serve to absorb the kinetic energy of the upstream jet and allow the flow to undergo a normal expansion. For a measuring plane after the settling screen, this screen can ensure a substantially uniform flow ahead of the measuring plane. The measured pressure difference between the static pressure at the fan outlet and the total pressure at the fan inlet is taken as the static pressure rise of the tested fan. The airflow rate can be calculated from the measured pressure drop across the nozzle(s). The selection of the nozzle is based on the criteria that the Reynolds number has to be higher than 12,000, so that the discharge coefficient of the nozzle becomes 0.95.

The operating point of the fan is controlled by adding a throttling device and an auxiliary fan at the end of the test chamber. The auxiliary fan is used to overcome pressure losses.

The thermometers are used to determine the air temperature in the duct and also to note the wet and dry bulb room temperature of air. The atmospheric pressure is determined and the density of air is calculated. The input power on the fan can be measured by recording the input voltages and currents.

8.6 Conclusion

In this chapter, the energy-efficiency expression for fans has been given. The measuring parameters which determine the energy-efficiency for fans have been pointed out. In the same principle, the volume flow rate, fan power and the static pressure rise are the key variables determining the energy-efficiency of fans. The energy-efficiency test rigs for electric motors, pumps and fans have been presented and discussed.

9

How to improve the energy-efficiency of electrical motor products

Abstract: The influence factors on the energy-efficiency of electric motors, pumps and fans are put forward in this chapter. Suggestions to improve the energy-efficiency of electrical motor products are presented.

Key words: energy-efficiency, improvement, electrical motor products.

9.1 Ways to improve the energy-efficiency of electric motors

9.1.1 Options for improving energy-efficiency in motors

Many options for improving energy-efficiency in motors are available (Prakash et al., 2008). These options include: use of high-efficiency motors; proper sizing of the motor to the load requirements (most motors operate at lower than the rated load because they were chosen in big capacity, performing inefficiently); use of adjustable speed drives, where appropriate, to match the speed and the torque to the load requirements (this allows in some cases the replacement of inefficient throttling devices and in other cases the simplification or even avoidance of wasteful mechanical transmissions); proper maintenance and repair (e.g., poor rewinding practices can damage motors and lower their efficiency significantly); and maintaining acceptable levels of power quality. Technology options for manufacturing high-efficiency electric motors are shown in Table 9.1 (Saidel et al., 2010).

Table 9.1 **Technology options for manufacturing high-efficiency electric motors**

Type of losses to reduce	Technology options applied
I^2R losses	Use copper die-case rotor cage
	Remove skew on conductor cage
	For improving thermic dissipation and cooling, increase cross-sectional area of rotor conductor bars; rotor bars are to be reshaped accordingly
	Increase end ring size
	Increased amount of copper in stator windings
	Manipulate stator slot size
	Improve the winding technique to reduce the length
Core losses	Improve grades of electrical steel
	Use thinner steel laminations
	Add stack height (i.e., length, add electrical steel laminations)
	Use high-efficiency lamination materials
	Use plastic bonded iron power
	Anneal steel laminations
Friction and windage losses	Use appropriate bearings and lubricant
	Install a more efficient cooling system
	Improve manufacturing tolerances
	Improve the aerodynamics of the motor
Stray-load losses	Narrower the radial air gap

9.1.2 Design of electric motors

Stator design

The stator is the outer body of the motor which houses the driven windings on an iron core. In a single-speed three-phase motor design, the standard stator has three windings, while a single-phase motor typically has two windings. The stator core is made up of a stack of round pre-punched laminations pressed into a frame, which may be made of aluminium or cast iron. The laminations are basically round with a round hole inside through which the rotor is positioned. The inner surface of the stator is made up of a number of deep slots or grooves around the stator. It is into these slots that the windings are positioned. The arrangement of the windings or coils within the stator determines the number of poles that the motor has.

A standard bar magnet has two poles, generally known as north and south. Likewise, an electromagnet also has a north and a south pole. As the induction motor stator is essentially like one or more electromagnets depending on the stator windings, it also has poles in multiples of two, i.e., two-pole, four-pole, six-pole, etc.

The number of poles determines the speed of the induction motor. A two-pole motor has a synchronous speed of 3,000rpm at 50Hz and 3,600rpm at 60Hz. A four-pole motor runs at half this speed, a six-pole motor at one-third speed and an eight-pole motor at one-quarter speed. The winding configuration, slot configuration and lamination steel all have an effect on the performance of the motor. The voltage rating of the motor is determined by the number of turns on the stator and the power rating of the motor is determined by the losses which comprise copper loss and iron loss, and the ability of the motor to dissipate the heat generated by these losses. The stator design determines the rated speed of the motor and most of the full-load, full-speed characteristics.

Rotor design

The rotor comprises a cylinder made up of round laminations pressed onto the motor shaft and a number of short-circuited windings. The rotor windings are made up of rotor bars passed through the rotor, from one end to the other, around the surface of the rotor. The bars protrude beyond the rotor and are connected together by a shorting ring at each end. The bars are usually made of aluminium or copper, but sometimes made of brass. The position relative to the surface of the rotor, shape, cross-sectional area and material of the bars determine the rotor characteristics. Essentially, the rotor windings exhibit inductance and resistance, and these characteristics can effectively be dependent on the frequency of the current flowing in the rotor.

A bar with a large cross-sectional area will exhibit a low resistance, while a bar of a small cross-sectional area will exhibit a high resistance. Likewise, a copper bar will have a low resistance compared with a brass bar of equal proportions.

Positioning the bar deeper into the rotor increases the amount of iron around the bar and consequently increases the inductance exhibited by the rotor. The impedance of the bar is made up of both resistance and inductance, and so two bars of equal dimensions will exhibit a different AC impedance depending on their position relative to the surface of the rotor. A thin bar which is inserted radially into the rotor, with one edge near the surface of the rotor and the other edge towards the shaft, will effectively change in resistance as the frequency of the current changes. This is because the AC impedance of the outer portion of the bar is lower than the inner impedance at high frequencies, lifting the effective impedance of the bar relative to the impedance of the bar at low frequencies where the impedance of both edges of the bar will be lower and almost equal. The rotor design determines the starting characteristics.

9.2 Control strategies for improving the efficiency of induction motors

A vast majority of motion systems are driven by induction motors. One of the advantages of induction (asynchronous) motors in comparison with synchronous ones is their capability to increase speed up to synchronous speed or decrease it to a fraction of their nominal speed (Dalvand and Zare, 2006). Historically, these are fixed-speed systems using mechanical mechanisms to control the flow, such as throttle or bypass techniques or even on-off operation, wasting enormous amounts of energy.

Many efforts have focused on the improvement of the materials, the design and the construction techniques to improve the efficiency of induction motors. However, in recent years, efficiency optimization control of motor drives has been the major research area for saving energy in motor applications. Efficiency-optimizing controllers find this value for each operating point and apply it to the machine under control. There are basically two methods for efficiency optimization control: off-line and on-line methods (Vaez-Zadeh and Hendi, 2005).

Keli Shi (2001) summarized the induction motor control methods, which include scalar control, vector control, speed sensorless control and intelligent control. Typical applications are scalar control, including open-loop voltage/frequency (V/Hz) control, closed-loop V/Hz control, and stator current and slip-frequency control. A variable speed drive is a kind of scalar control method. By introducing variable speed drive, consisting of a standard induction motor and a frequency converter, it is possible to optimize the efficiency of the electrical motor system. In Chapter 10, the control strategies for improving the efficiency of induction motors are addressed in detail.

9.3 Ways to improve the energy-efficiency of pumps

It is helpful to use a numerical simulation method for getting a better understanding of the complex flow phenomena. The use of CFD plays an important role in fluid mechanics. Numerical analysis based on three-dimensional Reynolds-averaged Navier–Stokes (RANS) equations has been widely used for the flow analysis of turbomachinery with the aid of the recent development of computing power. Application of this analysis has reduced the cost and time for the design of pumps by reducing the number of experimental tests. Based on experimental tests and numerical simulation, impeller inlet geometry has an important influence on performance improvement in the case of centrifugal pumps.

9.4 Ways to improve the energy-efficiency of fans

For the complexity of internal flow characteristics of fans, a set of wholly mature design calculation methods of impeller theory does not currently exist. The factors that affect efficiency of centrifugal fans are blade profile, wheel width, inlet-to-wheel clearance and cut-off height. For axial fans, the important factors are blade profile, tip clearance and swirl recovery. Some fan types have inherent advantages or disadvantages due to their configuration. As an example, a forward-curved centrifugal fan cannot achieve peak efficiencies close to those achieved by any backward-oriented blade design because the forward-curved blade guarantees separated flow (locally) downstream of the blade, and this separation causes a loss of energy. The backward-oriented designs rank in efficiency order according to the aerodynamic quality of the airflow passage through the impeller; curved blades are better than flat blades and airfoil shapes are better than single thickness. For axial-fan types, propeller fans suffer from larger tip clearances and the absence of turning vanes and tube-axial fans suffer from the lack of turning vanes.

For low energy demand it is imperative that the fan is working within the optimum efficiency range and is properly sized. These influential effects have not been – and cannot be – addressed in a product-oriented standard because they are dependent on the specific application.

9.5 Conclusion

To improve the energy-efficiency of electric motors, the rotor loss, stator loss and core loss can be reduced by proper design and control methods. By CFD simulation method, pumps and fans can be designed more scientifically. The energy-efficiency for electrical motor systems can be improved greatly by introducing variable speed drive.

10

Modern control methods for the induction motor

Abstract: This chapter introduces the necessities of speed control when operating induction motors to improve their efficiency. Development of the control method for electric motors is described. Mathematic modelling of the induction motor is provided. Finally, energy-saving control methods for induction motors are analysed.

Key words: control methods, speed control, energy-efficiency saving.

10.1 Why adjustable speed is needed for induction motors

In the past, AC induction motors were mainly used in constant-speed application, such as driving compressors, pumps, fans and mills, connected directly to a utility power line. If the pumps or fans need to change the fluid volumes, the AC induction motor cannot regulate automatically according to the actual load. If the valves are used to control the fluid volume, part of the energy is consumed in the valves and the system efficiency is reduced. Compared with DC motors, AC induction motors were thought to be cheap devices, but with expensive, complicated controllers.

As we know, induction motors have high efficiency when working close to their rated torque and speed. However, at light loads, no balance in between copper and iron losses results in considerable reduction in the efficiency. In order to improve the part-load efficiency and power factor, it is important to develop appropriate control schemes to make the motor excitation adjustment in accordance with load and speed.

In the first half of the twentieth century, about 80 per cent of constant-speed electric drives were used by AC motors; only 20 per cent of adjustable-speed electric drives were used by DC motors. In the latter half of the twentieth century, with the revolutionary advances in computer technology, power electronics, modern control and artificial intelligence, the adjustable-speed realized by AC motors has prevailed. Now the AC adjustable-speed substitute for the DC adjust-speed is a fact.

10.2 The development of modern control for electric motors

The induction motor drive features dynamic, recurrent and non-linear control. Despite the great efforts devoted to induction motor control, many of the theoretical results cannot be directly applied to practical systems due to the following difficulties: 1) complicated computations involved; 2) non-linearity of the systems; 3) unpredictable (load and noise) disturbances; and 4) uncertainty in the machine parameters. Different control algorithms are designed based on the control theory, principle and hardware. The previous theories of induction motor control are classified into two categories: scalar control, which adjusts the supply magnitude/frequency, and vector control, which adjusts the supply magnitude/phase. With the development of a control technique for the induction motor, control algorithms can be classified into four main categories, as shown in Table 10.1 (Chan and Shi, 2011).

Table 10.1 Various control approaches for induction motors

Control theory	Controller	Advantages/Disadvantages	Hardware
Scalar control	1. Open-loop voltage/frequency (V/Hz) control 2. Slip frequency and voltage control 3. Slip frequency and current control	Stable/performance dependent on machine parameters	Micro-processor based
Vector control	4. Direct FOC 5. Indirect FOC 6. Direct self-control (DSC) 7. Acceleration control	High performance, flux and torque are controlled, performance is dependent on machine parameters, integral error accumulation	DSP based
Speed sensorless control	8. Model reference adaptive system 9. Luenberger observer	No speed sensor, performance is dependent on machine parameters	DSP based
Intelligent control	10. Fuzzy control	Performance approximates that FOC, less dependent on machine parameters	Micro-processor based
	11. Neural-network control	High performance, robust structure	Neural device based
	12. Expert-system control	High performance, performance is independent of machine parameter variations, no integral error accumulation, performance is dependent on high precision speed sensor	DSP based and high precision encoder

10.2.1 Scalar control

Scalar control is based on the steady-state model of the motor. The control is due to the magnitude variation of the control variables only and disregards the coupling effect in the machine. For example, the voltage of a motor can be controlled to control the flux, and frequency or slip can be controlled to control torque. However, flux and torque are also functions of frequency and voltage respectively. This method is simple and easy to implement, but the inherent coupling effect (i.e., both torque and flux are functions of voltage or current and frequency) gives sluggish response and the system is prone to instability because of a high-order (fifth-order) system effect. As a result, the scalar control technique has poor dynamic performance. The scalar controller is usually used in low-cost and low-performance drives.

10.2.2 Vector control

In 1971, German engineer F. Blaschke proposed vector control for asynchronous motors for the first time. The advent of vector control was a milestone for the control of electric motors. In the last forty years, vector control technology has been widely applied. The objective of vector control is to simplify the induction motor control equations so that they resemble the equations for separately excited DC motors. Two vector control strategies were formulated during the late 1960s and decoupling of the stator currents was achieved either by the direct measurement of the flux or by estimation from the motor model. Both of these methods are based on the space phasor modelling of induction machines. The behaviour of an AC induction motor drive is described by three independent variables – the speed, the terminal frequency and the parameters of the motor and its power supply. The development of field-oriented control at the beginning of the 1970s made it feasible to control the induction motor as a separately excited DC motor.

In 1985, German professor M. Depenbroch, Japanese scholars Takahashi and Noguchi introduced the direct torque control (DTC) principle for electric motors. Nowadays this is an industrial standard for induction motor drives (Morari and Thiele, 2005). In this method, stator voltage vectors are selected according to the differences between the reference and actual torque and stator flux linkage. The DTC method is characterized by its simple implementation and a fast dynamic response. DTC is a close-loop control of stator flux and torque. Because the controlled variable is stator flux, DTC is not influenced by the rotor parameters. However, the torque ripple is large, which limits the system dynamic performance. At present, vector control and direct torque control are continuing to develop.

10.2.3 Speed sensorless control

The vector control principle provides great flexibility for induction motor drives. However, it is costly to implement because of the need for a shaft speed or shaft position encoder. Sensorless control has been introduced in the drives literature to provide an alternative, yet lower-cost, solution and the speed sensorless control of induction motors is a new research trend. To eliminate speed and position sensors, many speed and position estimation algorithms have been proposed recently. There are several techniques proposed in the literature to obtain speed information of motors instead of using the speed encoder. These can be broadly classified as:

- open-loop speed control with slip estimation;
- closed-loop control with speed estimation.

In open-loop speed control, the motor synchronous speed or frequency is regulated while the estimated slip frequency is used only to compensate the load changes. In closed-loop control, the speed of the motor is estimated and is used as a feedback signal for closed-loop speed regulation. Among the closed-loop control methods, the model reference adaptive system (MRAS) and the Luenberger observer are researched in much of the literature (Paladugu and Chowdhury, 2007).

10.2.4 Intelligent control

Vector control heavily relies on the mathematical model of the motor; the motor parameters may vary greatly during its operation, so this control technique presents a high sensitivity to the variations of the motor parameters. If the hysteresis controller is used in the direct torque control, although the mathematic model of the motor is not needed, the accurate motor parameters are needed when the stator flux and torque are estimated.

Intelligent control is a new kind of control method in the automatic control field, which includes expert-system control, fuzzy-logic control, neural-network control and genetic algorithm. It is not only based on artificial intelligence theory, but also on conventional control theory. Intelligent control does not rely on the mathematic model of the control object in which the control target is the actual control result. The uncertainty in the control process is considered adequately and fully. On the other hand, intelligent control is highly non-linear. For example, the fuzzy logic and rules reasons are one kind of mapping, which reflects the non-linear nature of the system.

10.3 Mathematical models of three-phase AC machines

10.3.1 State equation expressed by state variables: ω - i_s - ψ_r

The dynamics of an induction motor under the assumptions of equal mutual inductances and linear magnetic circuit are given by the fifth-order non-linear differential equation with two inputs and only three state variables (Jadot et al., 2000; Shi, 2001; Marino et al., 1991).

The dq equations for a three-phase induction motor in the synchronous rotating frame can be written as:

$$\left.\begin{aligned}
\frac{d\omega}{dt} &= \frac{n_p^2 L_m}{J L_r}\left(\psi_{rd} i_{sq} - \psi_{rq} i_{sd}\right) - \frac{n_p}{J} T_L \\
\frac{d\psi_{rd}}{dt} &= -\frac{R_r}{L_r}\psi_{rd} + (\omega_s - \omega)\psi_{rq} + \frac{R_r}{L_r} L_m i_{sd} \\
\frac{d\psi_{rq}}{dt} &= (\omega - \omega_s)\psi_{rd} - \frac{R_r}{L_r}\psi_{rq} + \frac{R_r}{L_r} L_m i_{sq} \\
\frac{di_{sd}}{dt} &= \frac{L_m R_r}{\sigma L_r^2 L_s}\psi_{rd} + \frac{L_m}{\sigma L_s L_r}\omega\psi_{rq} - \gamma^* i_{sd} + \omega_1 i_{sq} + \frac{1}{\sigma L_s} u_{sd} \\
\frac{di_{sq}}{dt} &= \frac{L_m R_r}{\sigma L_r^2 L_s}\psi_{rq} - \frac{L_m}{\sigma L_s L_r}\omega\psi_{rd} - \gamma^* i_{sq} - \omega_1 i_{sd} + \frac{1}{\sigma L_s} u_{sq}
\end{aligned}\right\} \quad (10.1)$$

In equation 10.1, rotor speed ω, rotor flux linkage (ψ_{rd}, ψ_{rq}), and stator current (i_{sd}, i_{sq}) are state variables, while rotor moment of inertia J, the stator, rotor and mutual inductances (L_s, L_r, L_m), stator and rotor resistances (R_s, R_r) and the number of poles are the motor parameters. The voltages

(u_{sd}, u_{sq}) are inputs and T_L is the load torque. The subscripts s and r stand for stator and rator; d and q denote the components of a vector with respect to a fixed stator reference frame and $\sigma = 1 - \frac{L_m^2}{L_s L_r}$ is the motor leakage inductance, $\gamma^* = \left(\frac{L_m^2 R_r + L_r^2 R_s}{\sigma L_s L_r^2} \right)$.

10.3.2 State equation expressed by state variables: ω - i_s - ψ_s

In the d-q reference frame, the state variables and input variables are selected as follows, respectively:

$$X = [\omega \quad \psi_{sd} \quad \psi_{sq} \quad i_{sd} \quad i_{sq}]^T \tag{10.2}$$

$$U = [u_{sd} \quad u_{sq} \quad \omega_1 \quad T_L]^T \tag{10.3}$$

The output variables are expressed as:

$$Y = [\omega \quad \psi_s]^T \tag{10.4}$$

Then the motor equation can be written in d-q coordinates fixed to the stator as:

$$\left.\begin{aligned}
\frac{d\omega}{dt} &= \frac{n_p^2 L_m}{J L_r}\left(\psi_{sd} i_{sq} - \psi_{sq} i_{sd}\right) - \frac{n_p}{J} T_L \\
\frac{d\psi_{sd}}{dt} &= -R_s i_{sd} + \omega_1 \psi_{sq} + u_{sd} \\
\frac{d\psi_{sq}}{dt} &= -R_s i_{sq} - \omega_1 \psi_{sd} + u_{sq} \\
\frac{di_{sd}}{dt} &= \frac{R_r}{\sigma L_r^2 L_s}\psi_{sd} + \frac{1}{\sigma L_s}\omega\psi_{sq} - \gamma^* i_{sd} + (\omega_1 - \omega) i_{sq} + \frac{1}{\sigma L_s} u_{sd} \\
\frac{di_{sq}}{dt} &= \frac{R_r}{\sigma L_r^2 L_s}\psi_{sq} - \frac{1}{\sigma L_s L_r}\omega\psi_{sd} - \gamma^* i_{sq} - (\omega_1 - \omega) i_{sd} + \frac{1}{\sigma L_s} u_{sq}
\end{aligned}\right\} \tag{10.5}$$

10.4 Speed control using variable voltage and variable frequency

Both voltage and frequency can be controlled to control the motor flux. Adjustable voltage and frequency is one kind of speed control method to change the rotation speed. When the number of poles of a motor is fixed, the speed of an induction motor is changed with the frequency, that is:

$$N = \frac{120 f_1}{n_p}(1-s) \tag{10.6}$$

Where n_1 is the speed, f_1 is the frequency of supply, n_p is the number of poles and s is the slip.

The air gap induced electric magnetic field (EMF) in an AC motor is given by:

$$E_g = 4.44\, f_1 N_s K_{Ns} \Phi_m \tag{10.7}$$

Where f_1 is the supply frequency, N_s is the number of turns per phase in the stator, K_{Ns} is the winding cording factor and Φ_m is the peak air gap flux. Neglecting the stator impedance (resistance and leakage inductance), the induced EMF approximately equals the supply phase voltage. Hence:

$$U_s \approx E_g = 4.44\, f_1 N_s K_{Ns} \Phi_m \tag{10.8}$$

As we can see from equation 10.4, if the frequency f_1 is constant, the air gap flux is approximately proportional to the supply voltage, $\Phi_m \propto E_g \approx U_s$. To maintain the flux constant, the ratio between the supply voltage and frequency has to remain constant.

Variable-voltage, constant-frequency (VVCF) control

Keeping the frequency f_1 equal to the rated frequency f_{1N}, the stator voltage is adjusted separately; this method is called variable-voltage speed control. The motor speed can be controlled by variation of the supply voltage at constant frequency. Note that a large range of speed control is possible if the machine has high starting torque. The disadvantages of this type of drive are higher copper loss and flow of large harmonic currents in line with poor power factor (Bose, 2006).

Variable-voltage, variable-frequency (VVVF) control

From equation 10.4, the air gap flux is proportional to U_s / f_1. The frequency control requires proportional control in applied voltage, because then the stator flux remains constant. Otherwise, if frequency alone is controlled, the flux will change. When frequency is increased, the flux will decrease and the torque developed by the motor will decrease. When frequency is decreased, the flux will increase and may lead to the saturation of magnetic circuit. When the speed is below the base speed, the motor runs at constant torque operation range. When the speed is above the base speed, the motor runs at constant power operation range (Trzynadlowski, 2001; Warne, 2005).

Slip frequency control method

Obviously, VVCF and VVVF control methods are open-loop control – the speed of the motor is not precisely controlled by this method because the frequency control only controls the synchronous speed (Ahmad, 2010). The scheme of speed control with slip frequency is shown in Figure 10.1 (Yi and Boshi, 2010). A speed encoder measures the speed of the motor ω_r. This value of speed is compared with the speed command ω^*. The speed loop error is used to generate the command value of slip ω_s^* through a PI

Figure 10.1 The scheme of speed control with slip frequency

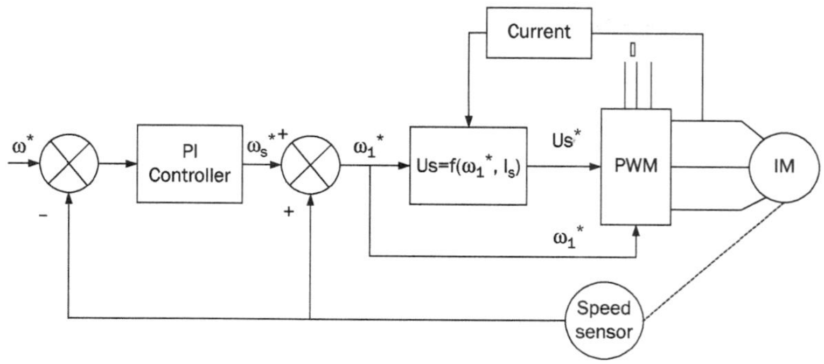

controller and limiter. The slip command is added to the measured speed ω to generate the command frequency signal ω_1^*. The voltage command signal U_s^* for the PWM inverter is also generated using frequency command ω_1^* and *V/f* function generator.

Power electronics for variable-voltage, variable-frequency inverter

According to Professor R.G. Hoft of the University Missouri-Columbia, *The Bibliography of Electronic Power Converters* (published by AIEE in February 1950) contains a chronological list of references, and lists that the first inverter paper was published in 1925 (Yano et al., n.d.). Earlier, when high power semiconductor devices were not available, it was a difficult task to obtain variable frequency supply. However, pulse width modulation (PWM) inverters are now available that can easily provide variable frequency supply with good-quality output wave shape. PWM technology enabled the elimination of harmonics from the inverter output voltage, allowing quasi-sinusoidal machine waveforms and eliminating torque pulsations.

10.5 Field-oriented control

Field-oriented control (FOC) was developed in the early 1970s and made it feasible to control the induction motor as a separately excited DC motor (Barambones and Alkorta, 2011). Field-oriented control can be classified as indirect field-oriented (IFO) and direct field-oriented (DFO) control. Both of these approaches provide torque control of the induction motor by decoupling the torque and flux (Nounou and Rehman, 2007). There are different ways of implementing the vector control strategy according to the choice of the reference frames for the space vectors. In an induction motor, there are three distinct flux space phasors: air gap flux, stator flux and rotor flux.

10.6 Direct torque control

Direct torque control (DTC) is different from the FOC scheme in the sense that the reference frame here is stator flux instead of rotor flux, which is used in the FOC scheme. The DTC control scheme abandons the stator current control philosophy: it directly controls the flux itself. Under the classic DTC scheme, if it is assumed that the rotor flux remains constant over the time interval of interest and the stator resistance drop does not exist, the stator voltage remains constant over the time interval. The stator flux ψ_s can be written as:

$$\psi_s(t_{K+1}) - \psi_s(t_K) = \Delta T u_s(t_K) \tag{10.9}$$

Equation 10.9 indicates that the stator flux always follows the direction of the voltage vector (Huangang et al., 2005).

The basic schematic of DTC is shown in Figure 10.2 (Vasudevan et al., 2006).

Figure 10.2 The open-loop DTC block diagram

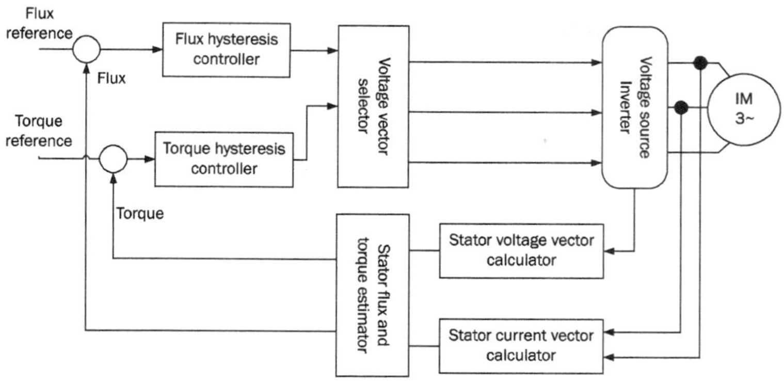

It achieves bang-bang torque and flux control by directly modifying the stator voltage in accordance with the torque and flux errors. The input to the direct torque controller are the torque error, error in magnitude of the stator flux space vector and the angle of the stator flux space vector, from which the states of the power switches are determined (Ahmad, 2010). In classical DTC, the switching frequency of the inverter is variable and is very low in comparison with the system sampling frequency. In addition, the inverter can only generate a restricted set of control signals for all operating conditions and therefore the maximum voltage of the DC link is always applied, which is independent of flux or torque error magnitude. These characteristics result in notable torque, flux and current ripples (Huangang et al., 2005).

10.7 Loss minimization control

In many electrical drives applications, both scalar and vector control techniques are usually operated by holding constant the amplitude of the rotor flux. Such an approach is pursued in order to obtain the best dynamic performance, while the drive efficiency is not a critical point because machine power losses are negligible (Cacciato et al., 2006). When the induction motor operates at light load, its efficiency is reduced because the iron losses increase dramatically (Ghozzi et al., 2004).The efficiency can be improved when the load torque is lower than the rated one, at the cost of slightly worse drive dynamic performance. If significant energy savings are to be obtained, it is necessary to optimize the efficiency of motor drive systems. To improve the motor efficiency, the flux must be reduced, obtaining a balance between copper and iron losses. It can be accepted in those applications where fast response requirements are not so strict. Several minimum-loss control schemes have been proposed in the literature. The control strategy of efficiency maximization techniques may be broadly divided into two categories (Lim and Nam, 2004). The first category is the so-called loss-model-based controller (LMC). The second category is the flux-search controller (FSC):

1. The loss-model-based controller (LMC) computes losses by using the machine model and selects a flux level that minimizes the losses. LMC is fast and does not produce torque ripple. The LMC approach is heavily dependent on parameter variation effects and simplifying assumptions made in the formulation of the loss function. These LMC drawbacks lead to a suboptimal flux setting.
2. The basic principle of the flux-search controller (FSC) is to measure the input power and then iteratively search for the flux level (or its equivalent variables) until the electrical input power settles down to the lowest value for a given torque and speed. Important drawbacks of

the search controller are slow convergence and torque ripples (Nam and Uddin, 2006).

A popular method is to keep the output power constant and searching for the operating point where the input power has a minimum. The operating point can be varied by adjusting the V/f ratio, application of fuzzy-logic control to adjust the magnetizing current or by adaptive control. The minimum-loss operating condition is obtained by iteratively changing the flux level in small steps until the minimum input power is detected. The main advantage of search control is that an *a priori* knowledge of motor parameters is not required. However, because the relationship between input power and flux near the minimum-loss point is fairly flat, the input power must be accurately measured to prevent oscillatory response in the control.

10.8 Conclusion

Strong demand for high-performance motor drives in industry has stimulated the research on induction motor control. The speed-control techniques for induction motors have been described in detail. Based on the scalar control principle, the variable-voltage, constant-frequency (VVCF) control, variable-voltage, variable-frequency (VVVF) control and slip frequency control methods have been analysed respectively. The field-oriented control method and DTC are simple and highly reliable schemes. If loss minimization is considered, two kinds of control method – LMC and FSC – can be applied.

References

Chapter 1

Almeida, A.T., Fonseca, P. and Bertoldi, P. (2003) 'Energy-efficient motor systems in the industrial and in the services sectors in the European Union: characterisation, potentials, barriers and policies'. *Energy*. Vol. 28, No. 7, pp. 673–690.

AMCA. (2010) *Fan industry – facing up to the energy challenges*. Arlington Heights, Ill. AMCA. Available at: *http://www.amca.org/UserFiles/file/AMCA_2010Mike%20 Brendel.pdf* (viewed 5 March 2011).

Kaya, D., Yagmur, E.A. and Yigit, K.S. (2008) 'Energy efficiency in pumps'. *Energy Conversion and Management*, Vol. 49, pp. 662–673.

United Nations Foundation. (2007) *Realizing the potential of energy efficiency: targets, policies, and measures for G8 countries*. Available at: *http://sefi.unep.org/fileadmin/media/sefi/docs/industry_reports/RPEE_Exec_Sum_-_Updated_for_Web.pdf* (viewed 4 March 2011).

World Pumps. (2009) *A brief history of pumps*. Press release. 23 March. World Pumps. Available at: *http://www.worldpumps.com/view/813/a-brief-history-of-pumps-/* (viewed 5 March 2011).

Chapter 2

Eskom. (2010) *Blow the lid on ineffective energy usage: efficiency in blower and fans*. Available at: *http://www.eskomidm.co.za/wp-content/themes/eskom/pdfs/Industrial/124538_Fans_Brochure.pdf* (viewed 5 March 2011).

European Commission. (2009) *Implementing Directive 2005/32/EC of the European Parliament and of the Council with regard to ecodesign requirements for electric motors*. ((EC) No 640/2009). Brussels: The Commission of the European Union.

Gulich, J.F. (2010) *Centrifugal pumps*. 2nd ed. Heidelberg: Springer, p. 39.

Holopainena, T.P., Tenhunenb, A. and Arkkio, A. (2005) 'Electromechanical interaction in rotordynamics of cage'. *Journal of Sound and Vibration*. Vol. 284, pp. 733–755.

Kaya, D. (2003) 'Experimental study on regaining the tangential velocity energy of axial flow pump'. *Energy Conversion and Management*. Vol. 44, No. 11, pp. 1817–1829.

Sujitjorn, S. and Areerak, K.-L. (2004) 'Numerical approach to loss minimization in an induction motor'. *Applied Energy*. Vol. 79, pp. 87–96.

United Nations Environment Program. (2006) *Electrical energy equipment: fans and blowers*. Available at: *http://www.energyefficiencyasia.org/docs/ee_modules/Chapter-Fans%20and%20Blowers.pdf* (viewed 5 March 2011).

US DOE. (2003) *Improving fan system performance: a sourcebook for industry*. Available at: *http://www.cedengineering.com/upload/Improving%20Fan%20System%20Performance.pdf* (viewed 5 March 2011).

Chapter 3

Mahlia, T.M.I., Masjuki, H.H. and Choudhury, I.A. (2002) 'Theory of energy efficiency standards and labels'. *Energy Conversion and Management*. Vol. 43, No. 6, pp. 743–761.

Wiel, S. and McMahon, J.E. (2003) 'Governments should implement energy-efficiency standards and labels – cautiously'. *Energy Policy*. Vol. 31, pp. 1403–1453.

Yanti, P.A.A. and Mahlia, T.M.I. (2009) 'Considerations for the selection of an applicable energy efficiency test procedure for electric motors in Malaysia: lessons for other developing countries'. *Energy Policy*. Vol. 37, pp. 3467–3474.

Chapter 4

Air Movement and Control Association. (2007) *Laboratory method of testing fans for aerodynamic performance rating*. AMCA Standard 210. Arlington Heights, Ill.: Air Movement and Control Association.

Europump. (2003) *European guide to pump efficiency for single stage centrifugal pumps*. Available at: *http://re.jrc.ec.europa.eu/energyefficiency/motorchallenge/pdf/EU_pumpguide_final.pdf* (viewed 5 March 2011).

National Electrical Manufacturers Association. (2009) *NEMA MG 1 Motors and Generators*. NEMA Standards Publication MG 1-2009, Revision 1. Rosslyn, Virginia: National Electrical Manufacturers Association.

Standardization Administration of China. (2006) *Limited values of energy efficiency and evaluating values of energy conservation of small and medium three-phase asynchronous motors*. GB 18613-2006. Beijing: Standardization Administration of China.

Standardization Administration of China. (2010) *Minimum allowable values of energy efficiency and values of efficiency grade for small-power motors*. Available at: *http://members.wto.org/crnattachments/2010/tbt/chn/10_2993_00_x.pdf* (viewed 5 March 2011).

Chapter 5

Angrisani, L., D'Apuzzo, M. and Moriello, R.S.L. (2006) 'Unscented transform: a powerful tool for measurement uncertainty evaluation'. *IEEE Transactions on Instrumentation and Measurement*. Vol. 55, No. 3, pp. 737–743.

Deldossi, L. and Zappa, D. (2009) 'ISO 5725 and GUM: comparison and comments'. *Accreditation and Quality Assurance*. Vol. 14, pp. 159–166.

Dunn, P.F. (2010) *Measurement and data analysis for engineering and science*. 2nd ed. New York: CRC Press.

Fink, K. (2004) *Knowledge potential measurement and uncertainty*. Wiesbaden: Deutscher Universitats-Verlag, pp. 189–191.

Fraden, J. (2005) *Handbook of modern sensors: physics, designs, and applications*. 3rd ed. New York: Springer, pp. 51–52.

International Organization for Standardization. (1993) *Guide to the expression of uncertainty in measurement.* ISO/IEC Guide 98:1993. Geneva: International Organization for Standardization.

Kessel, W. (2002) 'Measurement uncertainty according to ISO/BIPM-GUM'. *Thermochimica Acta*. Vol. 382, pp. 1–16.

Kimothi, S.K. (2002) *The uncertainty of measurements: physical and chemical metrology: impact and analysis*. Milwaukee: ASQ Quality Press, pp. 209–213.

Menditto, A., Patriarca, M. and Magnusson, B. (2007) 'Understanding the meaning of accuracy, trueness and precision'. *Accreditation and Quality Assurance*. Vol. 12, No. 1, pp. 45–47.

Pavese, F., Bar, M. and Forbes, A.B. (2009) *Advanced mathematical and computational tools in metrology and testing: AMCTM VIII*. Singapore: World Scientific, pp. 406–407.

Rabinovich, S.G. (1999) *Measurement errors and uncertainties: theory and practice*. 2nd ed. New York: Springer.

Wallace, A.K. and Wiedenbrug, E.J. (1999) 'Motor efficiency determination: from testing laboratory to plant installation'. *Pulp and Paper Industry Technical Conference*. pp. 190–195.

Wheeler, A.J. and Ganji, A.R. (2009) *Introduction to engineering experimentation*. 3rd ed. New Jersey: Pearson, p. 10.

Chapter 6

Agamloh, E.B. (2009) 'The repeatability of IEEE standard 112B induction motor efficiency tests'. *IEEE Electric Machines and Drives Conference*. pp. 1119–1126.

Aoulkadi, M. and Binder, A. (2006) 'Evaluation of different measurement methods to determine stray load losses in induction machines'. *International Symposium on Power Electronics, Electrical Drives, Automation and Motion*. pp. s13–s18.

IEEE Power Engineering Society. (2004) *IEEE Standard Test Procedure for Polyphase Induction Motors and Generators*. IEEE Std 112-2004. New York: Institute of Electrical and Electronics Engineers.

International Electrotechnical Commission. (2007) *Rotating electrical machines – Part 2-1: Standard methods for determining losses and efficiency from tests (excluding machines for traction vehicles)*. IEC 60034-2-1: 2007. Geneva: International Electrotechnical Commission.

Saidur, R. (2010) 'A review on electrical motors energy use and energy savings'. *Renewable and Sustainable Energy Reviews*. Vol. 14, pp. 877–898.

Standardization Administration of China. (2008) *Test procedures for single-phase induction motor*. GB 9651-2008. Beijing: Standardization Administration of China.

Wenping, C. (2009) 'Comparison of IEEE 112 and New IEC Standard 60034-2-1'. *IEEE Transactions on Energy Conversion*. Vol. 24, No. 3, p. 802.

Chapter 7

International Standardization Organization. (1999) *Rotodynamic pumps – Hydraulic performance acceptance tests – Grades 1 and 2*. ISO 9906:1999. Geneva: International Standardization Organization.

Shiels, S. (1998) 'Locating the greatest centrifugal pump energy savings'. *World Pumps*. pp. 56–59.

Standardization Administration of China. (2005) *Rotodynamic pumps – Hydraulic performance acceptance tests – Grades 1 and 2*. GB 3216-2005. Beijing: Standardization Administration of China.

Chapter 8

International Standardization Organization. (2008) *Industrial fans – performance testing using standardized airways*. ISO 5801:2008. Geneva: International Standardization Organization.

Lin, S.-C. and Huang, C.-L. (2002) 'An integrated experimental and numerical study of forward-curved centrifugal fan'. *Experimental Thermal and Fluid Science*. Vol. 26, pp. 421–434.

Chapter 9

Dalvand, H. and Zare, M. (2006) 'Evaluation of energy efficiency measures in Iranian industrial 3-phase electric motors'. *First International Power and Energy Conference*. Putrajaya, Malaysia.

Prakash, V., Baskar, S. and Sivakumar, S. (2008) 'A novel efficiency improvement measure in three-phase induction motors, its conservation potential and economic analysis'. *Energy for Sustainable Development*. Vol. XII, No. 2.

Saidel, M.A., Ramos, M.C.E.S. and Alves, S.S. (2010) 'Assessment and optimization of induction electric motors aiming energy efficiency in industrial applications'. *International Conference on Electrical Machines*. Rome.

Shi, K. (2001) *Intelligent control for an induction motor*. PhD thesis. Hong Kong: Hong Kong Polytechnic University.

Vaez-Zadeh, S. and Hendi, F. (2005) 'A continuous efficiency optimization controller for induction motor drives'. *Energy Conversion and Management*. Vol. 46, No. 5, pp. 701–713.

Chapter 10

Ahmad, M. (2010) *High performance AC drives and power systems*. London: Springer, pp. 47–96.

Barambones, O. and Alkorta, P. (2011) 'A robust vector control for induction motor drives with an adaptive sliding-mode control law'. *Journal of the Franklin Institute*. Vol. 348, No. 2, pp. 300–314.

Bose, B.K. (2006) *Power electronics and motor drives: advances and trends*. Burlington, VT: Academic Press.

Cacciato, M., Consoli, A., Scarcella, G., Scelba, G. and Testa, A. (2006) 'Efficiency optimization techniques via constant optimal slip control of induction motor drives'. *International Symposium on Power Electronics, Electrical Drives, Automation and Motion*. pp. S32–38.

Chan, T.-F. and Shi, K. (2011) *Applied intelligent control of induction motor drives*. Beijing: John Wiley & Sons (Asia) Pte Ltd.

Ghozzi, S., Jelassi, K. and Roboam, X. (2004) 'Energy optimization of induction motor drives'. *IEEE International Conference on Industrial Technology (ICIT)*. pp. 602–610.

Huangang, W., Wenli, X., Geng, Y. and Jian, L. (2005) 'Variable-structure torque control of induction motors using space vector modulation'. *Electrical Engineering*. Vol. 87, pp. 93–102.

Jadot, F., Martin, P. and Rouchon, P. (2000) 'Industrial sensorless control of induction motors'. In A. Isidori, F. Lamnabhi-Lagarrigue and W. Respondek (eds) *Nonlinear control in the year 2000: lecture notes in control and information sciences*. Vol. 258. London: Springer, pp. 535–543.

Kim, D.H. and Park, J.I. (2005) 'Loss minimization control of induction motor using GA-PSO'. In R. Khosla et al.

(eds) *Knowledge-based intelligent information and engineering systems: lecture notes in computer science*. Vol. 3682. Berlin: Springer, pp. 222–227.

Lim, S. and Nam, K. (2004) 'Loss-minimising control scheme for induction motors'. *IEE Proceedings – Electric Power Applications*. Vol. 151, No. 4, pp. 385–397.

Marino, R., Peresada, S. and Valigi, P. (1991) 'Adaptive nonlinear control of induction motors via extended matching'. In *Foundations of adaptive control: lecture notes in control and information sciences*. Berlin: Springer, pp. 435–454.

Morari, M. and Thiele, L. (eds) (2005) *HSCC 2005, LNCS 3414*. Berlin: Springer, pp. 274–290.

Nam, S.W. and Uddin, M.N. (2006) 'Model-based loss minimization control of an induction motor drive'. *IEEE ISIE*. pp. 2367–2372.

Nounou, H.N. and Rehman, H. (2007) 'Application of adaptive fuzzy control to AC machines'. *Applied Soft Computing*. Vol. 7, No. 3, pp. 899–907.

Paladugu, A. and Chowdhury, B.H. (2007) 'Sensorless control of inverter-fed induction motor drives'. *Electric Power Systems Research*. Vol. 77, Nos. 5–6, pp. 619–629.

Shi, K. (2001) *Intelligent control for an induction motor*. PhD thesis. Hong Kong: Hong Kong Polytechnic University.

Trzynadlowski, A.M. (2001) *Control of induction motors*. San Diego, CA: Academic Press.

Vasudevan, M., Arumugam, R. and Paramasivam, S. (2006) 'Development of torque and flux ripple minimization algorithm for direct torque control of induction motor drive'. *Electrical Engineering*. pp. 41–51.

Warne, D.F. (2005) *Newnes electrical power engineer's handbook*. 2nd ed. London: Newnes, p. 287.

Yano, M., Abe, S. and Ohno, E. (n.d.) *History of power electronics for motor drives in Japan*. Available at: http://www.ieeeghn.org/wiki/images/4/49/Yano2.pdf (viewed 5 March 2011).

Yi, Y. and Boshi, C. (2010) *Control systems of electric drives-motion control systems*. Beijing: China Machine Press, pp. 150–151.

Index

FSC
www.fsc.org
MIX
Papier | Fördert gute Waldnutzung
FSC® C083411

Kolibri 360 GmbH
Ferdinand-Jühlke-Straße 7
99095 Erfurt, Deutschland
produktsicherheit@kolibri360.de